Soothsayers in Dante's *Inferno*
with Their Heads Twisted Backward

Illustration of Canto XX of Dante's *Inferno* (*Divine Comedy*), showing Dante and his guide, Virgil, witnessing the punishment of ancient diviners for trying to see too far into the future. Their heads were twisted backward so they could only look into the past. (Priamo della Quercia, mid-fifteenth century)

FUTURE STORIES

FUTURE STORIES

WHAT'S NEXT?

DAVID CHRISTIAN

LITTLE, BROWN SPARK

New York • Boston • London

Little, Brown Spark
Hachette Book Group
1290 Avenue of the Americas, New York, NY 10104
littlebrownspark.com

First Edition: June 2022
Little Brown Spark is an imprint of Little, Brown and Company, a division of Hachette Book Group, Inc. The Little, Brown Spark name and logo are trademarks of Hachette Book Group, Inc.

The publisher is not responsible for websites (or their content) that are not owned by the publisher.

The Hachette Speakers Bureau provides a wide range of authors for speaking events. To find out more, go to hachettespeakersbureau.com or call (866) 376-6591.

ISBN 9780316497459
LCCN 2021950394

10 9 8 7 6 5 4 3 2 1

MRQ-T

Printed in Canada

I dedicate this book to my grandchildren,
Daniel, Evie Rose, and Sophia.
They are the future. May the future be good to them.

Contents

Contents

FUTURE
STORIES

Introduction

If you can look into the seeds of time,
And say which grain will grow and which will not,
Speak then to me . . .

— BANQUO TO THE THREE WITCHES,
SHAKESPEARE, *MACBETH*, ACT 1, SCENE 3

Open a creaky door in a haunted house and your spine will tin-
gle. Anything could appear. We open doors into the future
every moment of our lives. What lies behind them? How can we
prepare for the unknown when, as St. Paul writes, "we see
through a glass, darkly"?[1] This book is about the hidden face of
time, the parts that seem to lie in darkness because we haven't
been there yet. It's about how we try to imagine and prepare for
and deal with whatever lurks in the strange place we describe as
"the future."

Trying to make sense of the future can feel like clutching at
air. And yet, as airy as it seems, the future shapes an enormous
amount of our thinking and feeling and doing. So much anxiety
and effort, so much hope, and so much creativity are directed at
the future. Indeed, it may be that *most* of our thinking is actually
about possible futures. Most of the time we react to likely futures
on autopilot. This is everyday future thinking. It is familiar and

banal, and it is carried out by biological and neurological processes and algorithms that feel intuitive because they mostly work below consciousness. This is the future thinking we deploy when crossing a road and calculating if an oncoming semitrailer is going to hit us. We really face the mystery of the future when we set out in new directions, when a baby is born, when we face a sudden crisis, when we move to a new country or try to imagine the future of planet Earth. This is conscious future thinking. And once we start thinking consciously and carefully about it, we soon realize how weird the future is.

This book describes how philosophers and scientists and theologians have thought about the future. It discusses how other creatures, from bacteria to bats and baobab trees, deal with the same deep mystery using immensely sophisticated biochemical and neurological machinery. It explores the unique way in which our own species thinks collectively and often consciously about the future and tries to shape it. Finally, it describes some of the futures we can imagine today, for the next few decades, and in billions of years' time. We will end with speculations about the end of time.

An Everyday Mystery

We face the strange existential mystery of the future in every moment of our lives. There seem to be many possible futures. Then, in a flash, all but one disappear, and we are left with a single present. We must deal with the present quickly because soon it will be flash frozen into memory and history, where it will change only as much as a fossilized mammoth can twist and pivot inside an ice-age glacier. On the other side of every creaky door we know there is an endless, impatient crowd of other possible futures lined up and waiting, some banal, some trivial, some mysterious, and some transformative. And we don't know which we will meet.

The mysteries that surround the future are enchanting as well as terrifying; they give life much of its richness, beauty, exhilaration, and meaning — its zing! Do we really want to know what lies behind every door? Two thousand years ago, Cicero asked, "Had [Julius Caesar] foreseen that in the Senate, chosen in most part by himself . . . he would be put to death by most noble citizens, some of whom owed all that they had to him, and that he would fall to so low an estate that no friend — no, not even a slave — would approach his dead body, in what agony of soul would he have spent his life!"[2] Cicero knew Caesar and may have seen him knifed in the Senate on the ides (the fifteenth) of March 44 BCE. When Caesar died, Cicero was writing his great work on divination. So the example was vivid and strongly felt. Our ignorance of the future gives life much of its drama and excitement. And it gives us the freedom to choose — and the moral duty to choose thoughtfully.

A lot of the time, though, we really do want to glimpse what lies ahead. What clues do we have? When we journey to another country, we can talk to people who have been there, or we can travel with *Lonely Planet* guidebooks, as nineteenth-century Europeans traveled with their Baedekers. As a professional historian, I have traveled to the past in my imagination, using Baedekers built from the records and memoirs of those who lived in the past. I was not traveling blind. When we enter the future, though, we have no guides because no one has been there. Not a soul. As the philosopher of history R. G. Collingwood reminded us, the future leaves no documents.[3]

Not knowing is scary because the future really, really matters. After all, as the futurist Nicholas Rescher puts it, "that is where we are all going to spend the rest of our lives."[4] So we all look for guidance. Our minds are constantly scanning the world, looking for patterns and trends and signs and imagining possible futures, good and bad; we try to interpret messages in dreams, or in the stars, or in the warnings or promises of soothsayers or financial advisers. We ask parents or doctors or teachers. Modern governments ask economists and statisticians and scientists (and sometimes pay them a lot of money). And we engage in all this activity because, though the future leaves no documents, we do have some clues about what may be coming. And sometimes we can make forecasts with what Leibniz called "moral [i.e., near] certainty." The sun will rise tomorrow; I will die someday; the government will insist I pay taxes. I cannot say these things with "absolute certainty." But I can get close enough. What I cannot do is predict the future in detail, except in rare cases, such as solar eclipses. Unlike the past, which sparkles with details, the future is a foggy world of vague shapes moving in the gloaming.

The strangest thing is that our only clues about the future lie in the past. That's why living can feel like driving a race car while staring into the rearview mirror. No wonder we sometimes crash. Like the soothsayers in Dante's *Inferno*, who were punished by having their heads twisted backward, we enter the future looking back to the past. So it is ironic that historians, who spend their time studying the past, spend so little time thinking about the future. One aim of this book is to make the case for linking past thinking (aka "history") and future thinking, so that we can use the past more skillfully to illuminate possible futures.

Today, thoughtful future thinking is particularly important because we live at a turning point in our planet's history. In the past century, we humans have suddenly become so powerful that we hold the future of Earth and its frail cargo in our unsteady hands. What we do in the next fifty years will shape the future of the biosphere for thousands, perhaps millions, of years. And what we do will be informed both by how we think about likely futures and by the futures we try to build. A deeper understanding of what the future is, how we can prepare for it, and which futures are most likely — these are increasingly important forms of knowledge not just for experts but for every thinking citizen in today's world.

And yet, despite the strangeness of the future, the amount of thought we invest in possible futures, and the fundamental importance of careful future thinking, we rarely teach the general skills of future thinking in our schools and universities. We teach particular future-thinking skills such as computer modeling to specialists, but most of us have to wing it. We rely on our instincts and intuitions to cope with the mysterious world that lies just ahead of us and haunts so much of our thinking and so many of our actions. I have written this book partly because I realized how little I knew about what we mean by "the future" or about the subtle skills required to think carefully about likely futures. And yet I could not find any accessible general introductions to the future and future thinking.[5] I suspect I am not alone in wanting to know more about the strange world beyond the creaky door. So I have tried to write the book I was looking for. I think of it as a sort of *User's Guide to the Future.* Though not a specialist in future thinking, I have tried to make sense of what we *mean* by "the future," to understand better how to *think* about likely futures, and to use that understanding to *imagine* futures for ourselves, our planet, and the universe as a whole.

A Big-History Perspective

This book explores our thinking about possible futures through the multiple lenses of "big history," an emerging transdisciplinary field that has dominated my teaching and writing for three decades.[6] Big history views the past at all possible scales and from many different scholarly perspectives, with the belief that a sort of triangulation will yield a richer and deeper understanding of history. David Hume said that he found pleasure in taking a problem "pretty deep."[7] That, I hope, is what a big-history perspective can do for our understanding of the future. Imagine picking up a crystal that represents the future. In the chapters that follow, we will turn the crystal many times, looking at the future through different facets, in different lights, and through the eyes of experts of many different kinds. Each time we turn the crystal, its shape, color, and meaning will shift slightly, and we can learn something new.

Looking at a problem from multiple perspectives can be powerful. A fascinating theorem in network theory, known as the small-world theorem, suggests why. It shows that, in a network in which most points are neighbors, just one or two long-distance links can transform the entire network by accelerating the speed with which ideas and information and goods are exchanged. Most of human history has been shaped by village-size networks made up of neighbors with similar perspectives. But if just one villager travels regularly to the nearest town, they can revolutionize local networks by linking them to larger flows of information and very different perspectives. That is why small numbers of linkers — tramps, merchant caravans, peddlers, itinerant prophets, and soldiers — have played such a revolutionary role in human history. The ancient Silk Roads transformed Eurasian history by spinning webs of exchange — not just for goods, but for information and culture as well — from Korea to the Mediterranean.[8] Moving between scholarly disciplines can be as powerful. Discipline crossers created the basic paradigms of modern science, such as big bang cosmology, which linked the physics of the very large and the very small, or modern genetics, which links chemistry, biology, and physics. Like the Silk Roads, a big-history perspective can weave together threads from many domains of knowledge, creating new insights and ways of thinking. Forging new links can be particularly important in a field as difficult and fragmented as future thinking. As Wendell Bell, one of the pioneers of modern "futures studies," writes, "In a world of specialists and specialized knowledges, there is an important — and at present neglected — role to be played by the person who sees the big picture, who sees how different things interrelate, and who sees the whole and not only some of the parts."[9]

Of course, crossing disciplines is risky, like traveling the Silk Roads. There is a trade-off between local knowledge and the big picture. My hope is that insights from multiple perspectives will balance the occasional loss of graininess or nuance or precision. The quantum physicist Erwin Schrödinger articulates this dilemma well in the preface to *What Is Life?*, an interdisciplinary book that inspired Francis Crick and James Watson's breakthrough ideas on DNA. Aware that he was not a biologist, but convinced that physics had much to offer biology, Schrödinger wrote:

> I can see no other escape from this dilemma [the difficulty of linking insights from multiple disciplines] ... than that some of us should venture to embark on a synthesis of facts and theories, albeit with second-hand and incomplete knowledge of some of them — and at the risk of making fools of ourselves.[10]

This book explores the future in a similar spirit. It aims to take our understanding of the future "pretty deep," but it does so, paradoxically, by going pretty wide, by approaching the future from many different directions. It will explore how we try to *understand* the future, how we and other organisms try to *manage* different futures, how we humans try to *prepare for* the most likely futures, and finally, how we humans *imagine* the futures of our own species, our planet, and even our universe.

Future Stories' Origin Story

Why is a historian writing about the future? Most historians stick to the past. And quite rightly, according to R. G. Collingwood. "The historian's business," he thundered, "is to know the past not to know the future; and whenever historians claim to be able to determine the future in advance of its happening, we may know with certainty that something has gone wrong with their fundamental conception of history." Most historians agree. But Collingwood's argument is odd, because studying the past is the key to most forms of future thinking. And that is why not all agree with him. The historian E. H. Carr accepts that historians cannot predict specific events but insists that they can identify large historical patterns and trends that offer "general guides for future action which . . . are both valid and useful." Confucius would have agreed. "Study the past," he wrote, "if you would divine the future."[11] I hope to persuade some readers that historians may indeed have much to offer to future thinking.

It was big history that first persuaded me to think more seriously about the future. In the early 1990s, my colleagues and I at Macquarie University had just started the radical experiment of teaching a history course covering the whole of the past, from the dazzling moment, 13.8 billion years ago, when our universe was born in the big bang. Teaching such a course was ridiculously ambitious because it crossed so many traditional discipline borders. So it never occurred to us to discuss the future as well! Our final lecture was about today's world. After one such lecture, one of our best students approached me and said she loved the broad sweep of big history, "but" — I was waiting for that — "you can't stop in the present. If you've looked at fourteen billion years how can you *not* look at the next hundred years or so? How can you leave us on a cliff's edge? You have to talk about the future, too." I felt like banging myself on the forehead. Of course, she was right! The future is the rest of time, so shouldn't a historian spend some time thinking about it?

The next year, with a colleague, David Briscoe, who had given wonderful biology lectures in the course, we included a final lecture on the future. We didn't know what we were doing, of course. Which future? The next ten years? The next million? No idea. But David made a brilliant suggestion that ensured at least that the lecture would be fun. He said, let's not over-prepare. After all, *we really don't know what is going to happen!* So let's toss a coin in front of the students to decide who's the optimist and who's the pessimist. And let's take turns describing good futures and bad futures. So that's what we did. We agreed to have just one microphone so we would have to fight for it if we thought the other person was talking nonsense.

We gave these lectures for several years. Whatever else they were, they were fun. And they confirmed our student's instinct that if you try to think about all of the past, you cannot balk at thinking about the future. Yes, we experience past and future differently, and yet, like conjoined twins, they are inseparable. Thinking about the future led me to a rich, diverse, and some-times strange body of writing by theologians, philosophers, sci-entists, statisticians, science fiction writers, and scholars in futures studies.

Eventually, when I set out to write a history of the whole of the past, I followed my student's advice. The final chapter was about the future.[12] This book expands on that chapter. But it goes much further, as I have learned more about the future and realized how much of our thinking is actually about possible futures.

From here on, I will use the phrase *future thinking* to include all types of thinking about the future, even if they operate below consciousness. There are many other labels, from H. G. Wells's term *foresight,* to terms such as *futures studies, prognostics* (the term favored in the Soviet Union), to *planning, prediction* (often used critically for forecasts considered overconfident or over-precise), and *forecasting,* or the French word *prospective.* I will use the phrase *future management* to describe attempts, both conscious and unconscious, to control or steer the future in preferred directions.

Structure and Contents

This book is divided into four main parts and organized around four large questions.

The first part asks, "What is the future?" It describes what philosophers, scientists, and theologians have had to say about the future, and it describes the practical challenges all living organisms face as they try to *deal* with possible futures. Part 2 asks, "How do living organisms cope with the future?" It describes the sophisticated biochemical and neurological mechanisms that living organisms use as they try to manage uncertain futures. This is the foundation for all future thinking. Except in the brainiest of large organisms, these mechanisms work below the level of consciousness. Most future thinking takes place belowdecks. Part 3 focuses on the conscious future thinking of our own species. It asks, "How do human beings try to glimpse, understand, and prepare for the future?" Unlike the future thinking of other species, that of humans has changed radically since our species first appeared, so Part 3 describes human future thinking in three distinct eras of human history: the foundational era, up to about ten thousand years ago, the agrarian era, up to about two hundred years ago, and the modern era. Part 4 asks, "What sort of futures can we (plausibly) imagine for humanity, planet Earth, and the universe as a whole?" How should we imagine what may happen in the next hundred years or the next million, and . . . can we plausibly imagine the end of time?

PART I

Thinking about the Future

How Philosophers, Scientists, and Living Organisms Do It

CHAPTER 1

What Is the Future?

Time as a River and Time as a Map

We are placed in this world, as in a great theatre, where the true springs and causes of every event are entirely concealed from us; nor have we either sufficient wisdom to foresee, or power to prevent those ills, with which we are continually threatened. We hang in perpetual suspence [sic] between life and death, health and sickness, plenty and want, which are distributed amongst the human species by secret and unknown causes, whose operation is oft unexpected, and always unaccountable.
— David Hume, *Natural History of Religion*[1]

What is the future? The answer ought to be simple. After all, we live in time. So isn't the future just that part of time that hasn't happened yet?

The trouble is that once you start thinking hard about these questions, things get tricky very fast. There is not even a consensus about what the future is within modern futures studies. As Jim Dator writes, "'Time' and 'The Future' would seem to be two of the most central concepts for Futures studies, but in fact, 'time' was barely discussed by the founders of Futures studies, and has seldom been problematized subsequently."[2]

No wonder! Thinking about the future can make your brain hurt. The philosophy of time leads us into a scholarly jungle full of beautiful ideas, metaphysical thickets, and philosophical creepy-crawlies. I will try not to go too deep. But we have to venture in far enough to see the problems that gather, like lianas, around the ideas of time and the future.

To understand the future we need to understand time, but does time even exist? Or is the word just our name for a sort of conceptual ghost? In the humanities, some scholars prefer vaguer words such as *temporalities,* which can probably be translated as "experiences of temporal change."[3] Even modern science offers no complete answers. It's as if no one lives long enough to really get to grips with time. As Hector Berlioz (may have) said, "Time is a great teacher, but unfortunately it kills all its pupils."[4] Study time too much and, like the eleventh-century Persian astronomer and poet Omar Khayyám, you'll start to feel you're spinning like a Sufi dancer.

> *Myself when young did eagerly frequent*
> *Doctor and Saint, and heard great Argument*
> *About it and about: but evermore*
> *Came out by the same Door as in I went.*[5]

In Milton's *Paradise Lost,* even Satan's followers can't make sense of time.

> *[They] . . . apart sat on a hill retired*
> *In thoughts more elevate, and reason'd high*
> *Of providence, foreknowledge, will, and fate,*
> *Fix'd fate, free will, foreknowledge absolute;*
> *And found no end, in wand'ring mazes lost.*[6]

St. Augustine thought deeply about time as he searched for God's purpose. In the marvelous book 11 of his *Confessions,* which remains a fundamental text on time, Augustine asks: "What is time? Who can explain this easily and briefly?" Though he was a deep and subtle thinker, the problem of time always seemed to slip from his grasp. "What then is Time? Provided that no one asks me, I know. If I want to explain it to an inquirer, I do not know." So frustrated was Augustine that he begged God for help: "My mind is on fire to solve this very intricate enigma. Do not shut the door, Lord my God. Good Father, through Christ, I beg you, do not shut the door on my longing to understand these things which are both familiar and obscure." As the philosopher Jenann Ismael writes, "There is such a thing as too much deliberation."[7]

Two Approaches to Time

The problem of time has engaged philosophers, sages, farmers, shamans, theologians, logicians, anthropologists, biologists, mathematicians, physicists, gamblers, prophets, scientists, statisticians, poets, and soothsayers, as well as everyone worried about their own future and the futures of those close to them. Modern philosophers of time distinguish between two main approaches with very different implications for our understanding of the future.[8] Both are foreshadowed in ancient philosophical traditions. Heraclitus (fl. ca. 535–475 BCE) imagined a world of never-ending change. That meant that the future would be different from the past. His near contemporary, Parmenides, thought change was an illusion, so that past, present, and future should be much the same. Many philosophical and theological traditions have struggled with the relationship between permanence and change. The ancient Indian texts known as the Upanishads insist that there is "an inner core or soul (*atman*), immutable and identical amidst an outer region of impermanence and change." In many Buddhist traditions, though, "There is no inner and immutable core in things; everything is in flux."[9]

The first of our two metaphors follows Heraclitus. It sees time as a sort of flow, like a river, that carries us through never-ending changes. In this view, the future will be different from the past and hard to know. This is how we usually experience time in our everyday lives, so this metaphor feels natural for most of us today. This sort of time is like the turbulent world of ups and downs, joy and grief, birth and death, that some Indian traditions describe as samsara.

On the other side are those who argue that our sense of flow and change is a seductive illusion. "Real time," as a philosopher of time, the late D. H. Mellor, calls it, does not flow.[10] It is more like a map than a river. This approach is like a god's view of time, a view from above. From this perspective, change no longer looks like something that *happens* but more like the difference between two points on a map as experienced by an ant crawling between them. Our sense that the future is different from the past arises, in this view, from our own motion, not from the supposed flow of time. In this view, there is little difference between past and future, and in some sense the future should be knowable because it is already mapped. The idea that permanence lies beneath the superficial changes of everyday life may once have shaped most people's thinking about time, as I will argue in chapter 5. But in today's rapidly changing world, it is taken most seriously by philosophers and scientists worried by the logical puzzles thrown up by the idea of time as a flow, puzzles we will look at later in this chapter.

One metaphor implies that we are embedded in time, the other that we can perhaps stand above time. A recent survey of the philosophy of time refers to these two approaches as "dynamic" and "static" time. But philosophers often refer to them, in deference to a famous 1908 article by British philosopher J. Ellis McTaggart, as A-series time and B-series time.[11]

This is jargon, but the jargon is used so widely by philosophers of time that it is worth getting used to.

In practice, there is much overlap between the two metaphors. Even McTaggart, who saw time as an illusion, agreed that "we never *observe* time except as forming both these series."[12] We find a mash-up of these metaphors in one of the most famous definitions of time, that of Sir Isaac Newton. In *Principia Mathematica*, the most important work of the scientific revolution, Newton writes, "Absolute, true, and mathematical time, for itself, and from its own nature flows equably without regard to anything external, and by another name is called duration."[13] Newton's time "flows" like a river, but it is also "absolute," and it has extension, or "duration," like a line on a map.

Time as a River: The Future in A-Series Time

To make the metaphor of time as a river less abstract, let's join Mark Twain's young hero Huckleberry Finn and his friend Jim as they raft down the Mississippi:

> This second night we run between seven and eight hours, with a current that was making over four mile an hour. We catched fish and talked, and we took a swim now and then to keep off sleepiness. It was kind of solemn, drifting down the big, still river, laying on our backs looking up at the stars, and we didn't ever feel like talking loud, and it warn't often that we laughed — only a little kind of a low chuckle. We had mighty good weather as a general thing, and nothing ever happened to us at all — that night nor the next, nor the next.
>
> Every night we passed towns, some of them away up on black hillsides, nothing but just a shiny bed of lights; not a house could you see. The fifth night we passed

St. Louis, and it was like the whole world lit up. . . . Every night now I used to slip ashore toward ten o-clock at some little village, and buy ten or fifteen cents' worth of meal or bacon or other stuff to eat; and sometimes I lifted a chicken that warn't roosting comfortable, and took him along. . . . Mornings before daylight, I slipped into cornfields and borrowed a watermelon, or a mush-melon, or a punkin, or some new corn, or things of that kind.[14]

The flow of A-series time is as majestic as the Mississippi. It carries the flotsam of an entire universe, every star and galaxy, every atom and bug, into the future, just as the Mississippi carries rafts, fishing boats, canoes, yachts, paddle steamers, and driftwood. Our lives are part of that flow.

Huckleberry Finn and Jim live in a dynamic, ever-changing, Heraclitean world as their raft carries them into the future. Though some things seem similar, like the towns they pass each night, the details keep changing. Philosophers use the technical term *passage* to describe this feeling of never-ending change. The *Rubaiyat,* in Edward Fitzgerald's beautiful nineteenth-century translations of Omar Khayyám, captures the sense of *passage*:

> *Oh, come with old Khayyám, and leave the Wise*
> *To talk; one thing is certain, that Life flies;*
> *One thing is certain, and the Rest is Lies;*
> *The Flower that once has blown for ever dies.*[15]

The second thing we learn from the metaphor of time as a river is that the future lies in a particular direction. The raft carries its passengers downstream from their starting point in St. Petersburg, Missouri (Twain was probably thinking of his hometown, Hannibal). The future lies downstream, or ahead of us; or below us if, like many Mandarin speakers, you think of the past as up and the future as down; or behind us if, like many Australian Aboriginal communities and native Hawaiian speakers, you think of the future as behind your back.[16] Wherever it lurks, the future is in a different direction from the past.

The third thing we learn is that the future is hidden. At best we see a sort of fog, without the glittering details, the scents and colors, that give the past and present their iridescence. Huckleberry Finn can remember in the past "borrowing" a mushmelon or "lifting" a chicken that "warn't roosting comfortable." The present is fleeting, like the occasional "low chuckle" in the night. But while it's here, it is more real than anything. Only *now* can we feel the wind on our cheeks, or the pulse of a great river, or the heft of a "borrowed" watermelon, or the smell of a wood fire. So intense is our experience of the present that some philosophers ("presentists") argue it is the only reality. I remember hearing an English Buddhist monk, Ariyasilo, reminding us, "The past is gone. The future isn't here yet. Listen to the birds!"

In A-series time, past and future are very different. Figure 1.1 (on the following page) captures some of the differences. It was prepared by the Bank of England in 2013 to illustrate predictions about inflation. Sections before 2013 describe the past. They are based on detailed information and form a single line. After 2013, details vanish, and data points fan out into a misty cone of possibilities that is soon too wide to tell us anything useful. Just three years into the future, the Bank of England could only make the unhelpful prediction that 90 percent of likely outcomes fell within a range from a 0.5 percent decline in prices to a rise of almost 4.5 percent. Though separated only by the diaphanous veil of the *now,* past and future are very different in A-series time.

Particularly mysterious is the moment when past and future meet. As we raft downstream, it is as if we are approaching a ghostly supradimensional fleet of possible futures. But as they

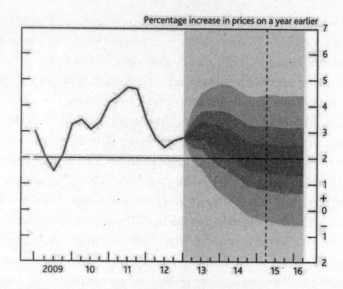

Percentage increase in prices on a year earlier

Figure 1.1: The Bank of England's May 2013 Fan Chart of Expected Inflation
The unshaded part shows inflation rates before 2013. They are known. To the right, we see one hundred likely inflation outcomes, assuming that "identical" conditions persisted as at the time the graph was created. The darker regions contain the outcomes thought to be most likely. The fan shows how quickly forecasts become too broad to say anything useful.
(From Kay and King, *Radical Uncertainty*, loc. 1625 Kindle.)

draw nearer, more and more of those possible futures pop into nothingness until, just as we reach them, the fog disperses and only one is left. The surviving future becomes a dazzling present before slipping into the past.

This is a bit like the strange process known to quantum physicists as the collapse of the wave function. The many possible positions and motions of millions of subatomic particles can be described mathematically by a probabilistic wave function that looks a bit like the Bank of England's predictions of future inflation rates. But if you measure the system, all the possibilities suddenly collapse into a single, detectable state, like the bank's descriptions of inflation rates in the past. In

A-series time, possible futures seem to collapse in a similar way as they reach us. Where did those other futures go? Did they ever really exist?

We can summarize the main features of A-series time in a type of diagram that we will return to several times in this book: a future cone.[17] To get an idea of the general shape of future cones, go back to Figure 1.1, which showed the Bank of England's predictions about future interest rates. Tidy the shape up, rotate it ninety degrees counterclockwise, and you have a diagram that includes the past and the future. It looks a bit like a cocktail glass because all our evidence suggests there is only one past, so the past appears as a single line while the future flares out into a cone with many possibilities.

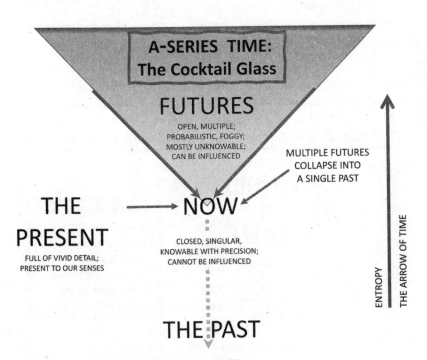

Figure 1.2: A-Series Time: The Cocktail Glass

Time as a Map: The Future in B-Series Time

A-series thinking about the future *feels* right to most people in today's world. But that was not always true. Philosophers of time, and traditional religions, know of a second type of time, which looks more like a map than a river. This is time as seen by the gods. McTaggart called this B-series time.

B-series time is simpler and more streamlined than A-series time. Past, present, and future are not so different from one another; they are just regions on a map. "Now" is where you happen to be at this moment, and the future is off to one side of your present position. Another observer will define present, past, and future differently, just as an observer in New York will imagine the West differently from an observer in Moscow. Here is a diagram capturing some features of B-series time. The first thing you may notice is that there is no cone! This diagram looks more like a worm than a cocktail glass.

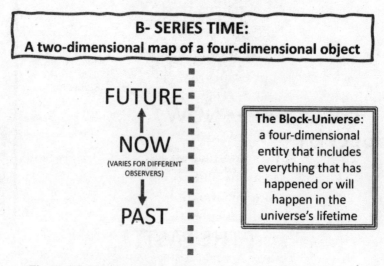

Figure 1.3: B-Series Time

Look at a to-do list or a school timetable and you are looking at the temporal equivalent of a map. Dentist, 9:45. Meeting, 11:30. Dinner with friends, 6:30. The schedule describes a temporal landscape within which future and past are just different places. And of course the metaphor of a map also suggests that the future is knowable: at 6:30 I will be meeting my friends.

B-series time adopts what Huw Price calls the "View from Nowhen," in which all moments are equal.[18] This is the view from above a map. Imagine flying high above the Mississippi and spotting Huckleberry Finn and Jim on their raft. Unlike them, you won't feel the chop of the current, but you can see where they have come from and where they are going. For you, different parts of their journey exist in a single space. If we could fly high enough, we might even imagine a map of everything that ever has existed or will exist in the universe. The coordinates of this universal map reach across all of space and time from the deepest past to the remotest future. Eventually, we are seeing a vast, frozen lump of all events and happenings and lives and deaths, the strange four-dimensional entity that the philosopher William James called the "block-universe." Einstein would call it the "space-time continuum." The block-universe is full of objects and events. The present moment is not that special because, as William James put it, "every event, irrespective of when it occurs, is fully and equally real, in the same way that events occurring at different *spatial* locations are fully and equally real."[19] Though he didn't have the modern jargon, St. Augustine seems to have believed that God saw a block-universe: "In the eternal, nothing is transient, but the whole is present." Or, as the philosopher Simon Blackburn puts it, "All events, past, present and future, exist like flies in amber, with greater or lesser distances between them."[20]

In the block-universe we shouldn't grieve for the dead or worry about the future. Albert Einstein captured this feeling in a letter of condolence to the family of his old friend Michele

Besso: "He has departed from this strange world a little ahead of me. That means nothing. For us believing physicists, the distinction between past, present and future is only a stubborn illusion."[21] The alien Tralfamadorians of Kurt Vonnegut's *Slaughterhouse Five* would have sympathized. They live in four dimensions so that, for them, no one dies because "all moments, past, present, and future, always have existed, always will exist." Similar views of time can be found within many philosophical and religious traditions. The thirteenth-century Japanese Zen monk Dogen writes, "Life is a position of time. Death is a position of time. They are like winter and spring, and in Buddhism we do not consider that winter *becomes* spring, or that spring *becomes* summer."[22]

B-series time has other odd features. With no definite "now" to anchor our images of reality, we have to think of everything as extended in time as well as in space, so we have to take more seriously the idea of time as a fourth dimension. That means that as I look down on Huck Finn and Jim, they may appear not as moving dots but as wormlike lines running down the river Mississippi. Kurt Vonnegut's Tralfamadorians see human beings as huge millipedes, "with babies' legs at one end and old people's legs at the other." The metaphor of a map also threatens our sense that change can occur in only one direction, from past to future. On a map you can move in all directions, so why not move back in time as well as forward?

Many philosophers and scientists are willing to live with the eccentricities of B-series time because A-series time seems to generate even more philosophical and logical conundrums. Take the idea of *now*, the moment that separates past from future. In B-series time, now is not special. It's just where/when you happen to be. But in A-series time, now is a special place that really is different from past and future. So shouldn't we be able to draw a line around now? How long does now last? Augustine argued that "the present occupies no space."[23] That idea leads to paradoxes that the Greek philosophers knew well. How can something happen if there is no time for it to happen in? The philosopher Zeno (495–425 BCE) invited us to think of an arrow in flight. In an infinitely small moment, it cannot cover any distance. So it must be at rest. Ditto for the next moment and the one before it. Therefore, the arrow cannot be moving. The idea of an infinitely small now doesn't seem to work philosophically or intuitively.

But what if now is *not* infinitely small? Perhaps time, like matter and energy, is granular. Does that get us off the hook? Perhaps there is a smallest possible atom of time, a *chronon*. Perhaps a chronon is the time light takes to cross the smallest possible length in space: about 10^{-35} meters long. Of course, our inner experience of now cannot possibly be that small. William James called the psychological now "the specious present." It probably lasts two or three seconds, as the mind organizes multiple sensations into a single picture of now, because our perceptions depend on neurological processes that edit and link information from many sensors and processors, interpolate where data is missing, and take time to do it all.[24] As we experience it, the border between now and the future is a blur of impressions, images, thoughts, and sounds. But if the present is not infinitely small, then some of it must stick into the future and some into the past like a temporal cocktail stick. Doesn't that make nonsense of the idea that future, present, and past are different? B-series time avoids these paradoxes because it doesn't treat now as special.

St. Augustine raised another difficulty with A-series time. Where are the past and future when we are in the present, as we always are in A-series time? "Do [they] exist," he asked, "in the sense that, when the present emerges from the Future, Time comes out of some secret store, and then recedes into some secret place when the past comes out of the present?"[25] We never actually experience alternative futures. We only ever meet a single future, and by the time it arrives it has turned into the present. So in what sense do alternative futures exist before we meet just one member of the delegation? Did the delegation ever exist? In B-series time, futures are just places on a map so these problems do not arise.

And here's another problem that takes us very deep. If time flows, how fast does it flow? Huckleberry Finn timed the Mississippi's flow past its banks at four miles an hour (6.4 kilometers per hour). Can we time time? Only if we know what it is flowing past. Newton understood the difficulty and tried to solve it by distinguishing absolute time, which he saw as an ultimate framework, like the banks of the Mississippi, from relative time. Newton explained the idea of absolute time by turning to theology, a subject he thought about as deeply as physics. He argued that God's universal presence provides the ultimate grid for space and time. Though he later retracted the idea, he once described the universe, in a revealing metaphor, as "the Sensorium of a Being incorporeal, living and intelligent."[26]

In the secular world of modern science, theological solutions no longer work. Nineteenth-century scientists tried to replace Newton's idea of God as the ultimate grid of reality with the concept of an "ether," a gossamer-thin medium through which all energy and matter traveled, and against which you could measure their speed. There were many attempts to detect the ether, but none succeeded. The most famous attempt was the Michelson-Morley experiment, conducted in 1887. This assumed that the speed of light ought to be slower when traveling against or across the ether, so there ought to be a difference in the speed of two light beams that travel at ninety degrees to each other. But no difference could be detected. That left advocates of A-series time with a flow from past to future, but nothing against which the flow could be measured. In chapter 2, we will look at Einstein's revolutionary solution to this puzzle.

Determinism, Causation, and the Arrow of Time

B-series time avoids the paradoxes of A-series time, but it also raises two profound problems for future thinking. First, the idea of a block-universe can be interpreted to mean that the future is all stitched up, so there are no choices to be made. That seems to be the end of free will, ethics, and morality. Second, in B-series time, change seems to have no clear direction. That is a big problem for future thinking because it deprives us of one of the most powerful ways of forecasting the future: the idea that if A causes B, then when A happens, we can predict B in the near future. Kick a ball now, and I predict it will move in the near future. These questions about determinism and causation threaten fundamental assumptions about how to cope with the future. That's a high price to pay for the simplicity of B-series time.

Fortunately, there are good answers to these questions that preserve our intuitive sense that (1) we can shape the future because it is not completely predetermined by the past, and (2) causes precede effects because many forms of change occur only in one direction: from past to future.

Some of these arguments are old, but in their modern forms they depend on a fundamental shift in scientific thinking that occurred in the late nineteenth century, a change in how modern science and philosophy conceive of reality and the future. From the seventeenth century to the early twentieth century, most scientists found the idea of determinism both logical and inspiring. They hoped that science would discover more and more mechanical laws that would increase our ability to predict the future. They accepted that all events in a mechanical universe, from the death of the sun to me having an extra cup of coffee this morning, were / are / will be predetermined from the

moment of creation. Omar Khayyám captures the idea of determinism poetically:

> *With Earth's first Clay They did the Last Man knead,*
> *And there of the Last Harvest sow'd the Seed:*
> *And the first Morning of Creation wrote*
> *What the Last Dawn of Reckoning shall read.*[27]

If Omar Khayyám is right, the whole idea of planning for
possible futures is nonsensical. The game is fixed. Does B-
series time abolish the idea of choice, along with all our ideas
about responsibility, ethics, and morality? The answer is . . . not
necessarily.

The classic modern account of determinism comes from the
work of the great French scientist Pierre-Simon de Laplace.
Laplace was a brilliant mathematician and lived in an era of
buoyant confidence in the power of science. In 1814, he drew
out the deterministic logic of post-Newtonian science in a work
called "A Philosophical Essay on Probabilities."

> Present events are connected with preceding ones by a
> tie based upon the evident principle that a thing can-
> not occur without a cause which produces it. . . . We
> ought then to regard the present state of the universe
> as the effect of its anterior state and as the cause of
> the one which is to follow. Given for one instance an
> intelligence which could comprehend all the forces by
> which nature is animated and the respective situation
> of the beings who compose it — an intelligence suffi-
> ciently vast to submit these data to analysis — it would
> embrace in the same formula the movements of the
> greatest bodies of the universe and those of the lightest
> atom; for it nothing would be uncertain, and the
> Future, as the past, would be present to its eyes.

In practice, Laplace admitted, human understanding will always remain "infinitely removed" from the understanding of such an all-knowing being.[28] Our ignorance will preserve the illusion of free choice. But, he argued, free choice *is* an illusion.

The argument is old. Two thousand years ago, Cicero put it into the mouth of his brother, Quintus, in his Socratic dialogue *On Divination*. Quintus defends the Stoic argument that "all things happen by Fate" because there is "an orderly succession of causes wherein cause is linked to cause and each cause of itself produces an effect." From this Quintus concludes, like Laplace, that if we knew enough, we could predict the future. "The evolution of time is like the unwinding of a cable: it creates nothing new and only unfolds each event in its order."[29]

Extreme determinism has always worried theologians and philosophers because if humans have no freedom to choose, they cannot be held responsible for what they do, and that is the end of ethics and morality. For theologians in the Abrahamic tradition, the problem was how to square the idea of human freedom of choice with the idea of an omnipotent and omniscient God. For scientists, the problem is to decide whether scientific laws leave any wiggle room for individual choices or contingency or randomness.

There have always been powerful arguments against extreme determinism. In a polemic with Cicero, Augustine argued that we *do* have freedom of choice despite God's omnipotence and omniscience. God has granted us limited freedom to choose, but, with infinite "foreknowledge" and the ability to stand outside of time, he also "foreknew" what we would freely will![30] Modern philosophers of time offer analogous arguments. The block-universe is real, they argue. But it is constructed both by mechanical causes that are broadly predictable in principle, and by events that were unpredictable at the moment when they occurred, such as quantum events or the choices made by purposeful beings. The block-universe can be "seen" only by entities standing outside the flow of time, but it is constructed, at least in part, by entities embedded in that flow. Today, the idea that free will and determinism are compatible is known, unoriginally, as *compatibilism*.

Since Laplace's time, the tide has turned against extreme determinism even within the hardest of sciences, such as physics. Philosopher of science Harry Laudan writes that in the late nineteenth century most scientists abandoned hopes for complete certainty. Instead, they committed to "a more modest program of producing theories that were plausible, probable, or well tested. As Pierce and Dewey have argued, this shift represents one of the great watersheds in the history of scientific philosophy: the abandonment of the quest for certainty."[31]

There were several reasons for this profound shift in scientific ideas about knowledge, reality, and the future.

Philosophers showed that no logical system could guarantee certainty. Bertrand Russell offered as an example the apparently simple statement: "This statement is false." If it is false it cannot be true. If true, it must be false. In the 1930s, Kurt Gödel's "incompleteness theorems" showed that all logical systems must contain assertions that cannot be proven to be true. In computing, Alan Turing proved that it is impossible to determine in advance where all computer programs will lead.[32] More recently, the Swiss physicist and mathematician Nicolas Gisin has shown that even in the world of numbers, absolute precision may be unattainable.[33]

Early in the twentieth century, quantum physics undermined determinism in physics by showing that at subatomic scales, many events are inherently unpredictable. Shine a light at a surface with two holes in it and try to predict which hole a given photon will go through. It can't be done. That means, as the physicist Richard Feynman put it, that "the future . . . is unpredictable."[34] Really! Not just because of our ignorance. Today, these uncertainties haunt all of physics. Given that our universe consists of unpredictable subatomic particles organized into gazillions of different formations, these are powerful arguments against Laplace's extreme determinism. General laws and tendencies there certainly are, but they cannot determine the future in detail, so they do not allow perfect prediction even in principle.

Chaos theory offers another reason for abandoning hopes of perfect prediction. In the early 1960s, the meteorologist Edward Lorenz found that trivial differences in starting conditions can cascade through complex systems such as the weather to produce wildly different outcomes. What may seem like infinitesimally small initial differences can be magnified many times over by positive feedback loops. This is known as the "butterfly effect," after Lorenz's metaphorical idea that the flapping of a butterfly's wings somewhere on Earth could be amplified into a hurricane somewhere else. The COVID-19 pandemic is an example of a world-changing event caused by alterations to the genome of a single virus so small you would need an electron microscope to see it.

One of the most powerful arguments against strict determinism comes from evolutionary biology. If the future were determined with absolute precision, why would evolutionary processes generate so many entities (including ourselves) that seem to try to *intervene* in events? Why invest so much evolutionary energy in choice-making mechanisms if there are no choices to make? (We will discuss some of these choice-making mechanisms in later chapters.) This argument, too, has ancient roots. In Boethius's *Consolation of Philosophy*, written fifteen hundred years ago while Boethius was imprisoned in Pavia, Lady Philosophy asks whether the outcome of a chariot race might be predetermined. Boethius's response is that it could not be because "then the exercise of skill would be beside the point."[35] Exactly. Why would God give humans the ability to make skillful choices if he had fixed the race?

In summary, most modern interpretations of the universe agree that specific events and outcomes are not completely predetermined. As the physicist Phil Anderson put it in 1972, "The ability to reduce everything to simple fundamental laws does not imply the ability to start from those laws and reconstruct the universe." There is a little "give" in the universe of modern science. William James writes, "The parts have a certain amount of loose play on one another."[36] If Huckleberry Finn and Jim dip an oar into the Mississippi, they *will* be able to change course slightly. B-series time does not commit us to extreme determinism because there seems to be some wriggling going on inside the block-universe. Phew!

But we're still left with the problem of causation because B-series time seems as if it should allow change to occur both backward and forward in time, while the idea of causation requires change to occur in just one direction, in which causes precede effects.

By the early twentieth century, physicists understood that most fundamental physical equations seem to work equally well whether we imagine time moving forward or backward. Film electrons in motion and try to figure out if the film is running forward or backward. You can't do it.[37] Today, physicists working at research facilities such as the Large Hadron Collider outside Geneva routinely encounter particles, such as positrons, that seem to travel backward in time. For the fundamental particles of physics, there seems to be no direction to time.

That upends all our ideas about causation. Some welcomed this shift because the idea of causation was already in trouble. In the eighteenth century, David Hume showed that you can never catch causation red-handed. You can show that two events seem to be correlated. When you kick a ball, the ball will move away from you. But proving that the kick *caused* the ball to move is impossible. The problem is that there are so many possible causes. Did the contraction of muscles in my leg cause the ball to move? Or the absence of something locking the ball in place? Or the neurons in my brain that made me kick the ball? Or the big bang that created me, the ball, and the football pitch? As Bertrand Russell argued in 1912, the idea of causation leads to an infinite regress. Statisticians are familiar with the problem of hidden causes. In the 1950s, evidence accumulated for a correlation between smoking and lung cancer, but the British statistician Ronald Fisher, a notorious contrarian and smoker (and paid consultant to tobacco companies), argued that perhaps there was an undiscovered gene for both smoking and lung cancer, or perhaps lung cancer caused smoking! Such arguments are surprisingly hard to refute.[38]

So profound are these difficulties that in the early twentieth century, many, including Russell, suggested that science and philosophy should dump the idea of causation, along with the idea that time has a direction.[39] But even Russell hesitated, and with good reason — like many early-twentieth-century scientists, he was beginning to give up on the perfectly determined world of Newtonian science. And that led him to consider looser and more probabilistic ways of understanding causation and the relationship between past and future.

Even Hume conceded that, despite the logical difficulties it raises, the idea of causation is indispensable in practice because it works so well so much of the time. Russell agreed. It makes sense to talk of causal laws, as long as we do not regard them as "universal or necessary." In other words, we can use the idea of causation to forecast likelihoods with great confidence, even if absolute certainty is unattainable. "If . . . we know of a large number of cases in which A is followed by B, and few or none in which the sequence fails, we shall in *practice* be justified in saying 'A causes B,' provided we do not attach to the notion of cause any of the metaphysical superstitions that have gathered about the word."[40]

In the late twentieth century, the idea of causation reappeared in more modest forms. The computer scientist Judea Pearl showed that we can get rid of the infinite regress of causes if we think of causation from the perspective of local actors intervening in local processes.[41] That, after all, is the perspective from which we humans approach causation in the real world. We don't try to include all causes, but only those that seem to make a difference. What will happen right now if I kick this ball? I can make a pretty good prediction based on how hard I intend to kick it, whether it's inflated, whether it's locked in place, and so forth. Pearl has shown that this more modest approach to causation can be used with great mathematical rigor.

The idea of an arrow of time — the idea that time has a direction — also returned in more modest, perspectival, and probabilistic forms. When dealing with simple entities, such as subatomic particles, it is indeed hard to ascribe a direction to time. But in our everyday lives we deal with more complex structures and there we find plenty of evidence for an arrow of time. Film an egg being broken and scrambled and you will know the direction of time.[42] Time moves in the direction in which organized things get *less* organized, the direction in which the shell breaks, and the yolk and white get mixed together, *not* the direction in which a scrambled egg unscrambles itself.

Scientists describe all this in the technical language of thermodynamics, and thermodynamics is subtle stuff. They say that the "entropy," or disorder of energy and matter, tends to increase as you move from past to future. So, though the total amount of energy in the universe is fixed, energy tends to exist in less ordered forms as time passes. It becomes less like the ordered flows of an electric current and more like the random twitchings of heat energy, which, in extreme forms, is too chaotic to do much. More organized flows of energy ("free energy") can get more done; they can even arrange matter in more organized structures. But as free energy does work it becomes more ragged and less ordered, just as a battery eventually runs down. Entropy increases. This inexorable degrading of free energy gives a direction to all change. It ensures that energy will flow, and flows of free energy can build and sustain complex entities. But those same complex entities (including you and me) tend to discombobulate flows of energy as they use them so that, paradoxically, their existence accelerates the degradation of free energy.[43] As free energy is degraded, it will become harder for complex entities to exist, so that both energy and matter will tend to become less ordered. This is the idea behind one of the most fundamental of all scientific laws: the second law of thermodynamics.

Strictly speaking, the second law is not a law but a very powerful directional tendency in the evolution of our universe. There is no scientific law that prevents all the atoms of a scrambled egg from unscrambling themselves and inserting themselves back in a perfect shell. It's just that the odds against that happening randomly are extremely (colossally! staggeringly! mind-bogglingly!) high. Complex structures break down eventually because there are many more possible disordered arrangements than ordered arrangements, so if you keep spinning some cosmic roulette wheel, you can more or less guarantee you will end up with less ordered states. In short, we expect as a general rule (another "moral" certainty) that, unless there is some special input of more ordered "free energy" from the outside (unless someone tidies things up), complex structures will tend to get less complex as we move from the past to the future. In the future, your bedroom will get less tidy unless you clean it. The arrow of time points in the direction of increasing disorder and eventual breakdown.

There are other reasons for thinking that most change is directional. Drop a stone in a lake and the ripples will always move away from the center. They will never move inward. This is characteristic of any wavelike motion, including the movement of energy through the universe, for reasons that we do not fully understand.[44] But the most powerful example of a temporal direction can be found within big bang cosmology. Our universe is expanding in one temporal direction, toward the future.

Though B-series time does not rule out processes that move backward in time, it seems that large complex lumbering entities such as ourselves can ignore those possibilities as we focus on the task of coping with time and the future here on planet Earth. We can assume time has a direction so that, even in B-series time, we can use the idea of causation to help predict what is likely to happen in the future. Phew again!

In summary, though the block-universe of B-series time may seem to undermine our ideas about choice and causation, modern science tells us that, even in a block-universe, not everything is fixed in advance and most of the changes that affect us are directional so that causes really are followed by consequences. That means we really can make some choices about the future, and a lot of the time we really can rely on the idea of causation to forecast likely futures. Future thinking is possible! That's a relief!

CHAPTER 2

Practical Future Thinking

Time as a Relationship

The reconciliation of time as conceived in physics with time as encountered in experience is the central problem in the metaphysics of time.

— JENANN ISMAEL, "TEMPORAL EXPERIENCE"[1]

Chapter 1 explored some of the mysteries of the future through the philosophy of time. But in our everyday lives the future is no abstraction. What matters to us is not the *idea* of the future, however precise or rigorous that idea might be. What matters is what we must actually do, from moment to moment. We expect our ideas about the future to do some heavy lifting for us. If it could engage in philosophical debate, the tiniest of bacteria would probably agree with the last of Marx's "Theses on Feuerbach": "The philosophers have only interpreted the world, in various ways; the point is to change it."[2] So how, in practice, do we deal with the uncertainty? We live in the turbulence and darkness of A-series time, but we yearn for the knowable futures mapped out in B-series time. The great Indian religious epic, the Bhagavad Gita, or "Song of God," contains a core story that can be interpreted as a poetic exploration of this deep yearning.

The warrior prince, Arjuna, is about to enter battle. The "conch horns and kettledrums, the cymbals, drums, bull-mouth trumpets," have all sounded. He fears the future will bring terrible fratricidal slaughter because he sees in the opposing army "teachers, mother's brothers, brothers, sons, grandsons, and friends, too." He faces all the turbulence and terror of futures in A-series time, and he is appalled and confused. So he asks his charioteer, the god Krishna, to stop time's flow and create a sort of cosmic time-out. Krishna obliges, "[making] the great chariot stand between both armies." Arjuna and Krishna are now in a strange temporal border zone, free of the dynamism and specificity of A-series time, but without the godlike perspectives of B-series time. And in this still place, the prince asks the god for advice about the future. Arjuna is so appalled by the thought of the coming battle that he has decided not to fight. He "let go of both his bow and arrow, his whole being recoiling in grief." But Krishna explains that no one can avoid the battles of life: "No one, not even for one moment, ever stands without acting." Even inaction is action. Then Krishna offers Arjuna a tiny glimpse into his own god's-eye view of time, in which all futures are already mapped out. "I have come forth to destroy the worlds. Even without you, these warriors facing off against each other will no longer exist." In the block-universe that Arjuna glimpses through Krishna's eyes, there is no point in mourning his own death or that of his enemies. "I have never not existed," says Krishna, "nor have you, nor have these lords of men. Nor will we cease to exist."[3] Arjuna's brief glimpse into the unchanging realm of B-series time gives him the serenity he needs to act in the world. "Do not be troubled," Krishna tells him, "but strike!"

Like Arjuna, we all prepare for the future from within a particular place in time and space, but to act we need a wider and more universal vision of what is going on. So all future thinking is relational. It is a sort of negotiation between who and where we are right now and a wider universe that we struggle to see.

That means there is no one answer to the question: "What is the future and how does it work?" How you cope with the future depends on who you are and where and when you are in the universe.

Relativity and the Future

Early in the twentieth century, Albert Einstein showed with scientific precision the relational and perspectival nature of our experience of time. His remarkable paper on special relativity, which he published in 1905 as a twenty-six-year-old patent clerk in Bern, upended Newton's ideas on absolute time and transformed our understanding of time and the future.[4]

·It showed that there is no universal, absolute temporal flow. Instead, the rate at which time flows varies from observer to observer according to strict rules that depend on each observer's "frame of reference," their position and motion within the universe. As the German sociologist Norbert Elias wrote in a pioneering history of our changing experiences of time, "It was Einstein who finally set the seal on the discovery that time was a form of relationship and not, as Newton believed, an objective flow."[5]

Einstein's arguments start from the remarkable fact, well-known by 1900, that the speed of light seems to be absolute. That is very odd. If you travel toward the sun and measure how fast a sunbeam is traveling toward you, you will get exactly the same result as an observer who is traveling *away* from the sun or at right angles to it. All your speed guns will show that the beam is traveling at about three hundred thousand kilometers (186,000 miles) per second.[6] That's not how we expect things to work on Earth. If I measure the speed of a car coming toward me, I expect it to be different from the speed of a car moving away from me. Most of Einstein's contemporaries expected that these anomalies would get resolved: perhaps they were the result of experimental error. Einstein took a different approach. Perhaps the speed of light really was absolute, as all the experiments suggested, in which case it had to be the rulers and clocks and speed guns of different observers that were behaving oddly in order to always come up with the same result. "If," he wrote, "observers in different states of motion always find this same value [for the speed of light] then their measurements of space and time must differ." He explored this speculation with a famous thought experiment based on the fastest technology of his time: the railway.[7] If Einstein had developed his theory today, his metaphors would surely have included jet planes or rocket ships.

Here is a slightly modified version of his thought experiment. Imagine Isaac is standing at a train station and sees two simultaneous flashes of lightning. One is ten kilometers to the east and one is ten kilometers to the west. Of course, he knows that the flashes actually occurred slightly before he perceived them because it takes light time to travel. Now imagine Albert is on a train traveling eastward through the station at exactly the moment when the two lightning flashes occur. Will Albert agree with Isaac that the flashes were simultaneous? Einstein's answer is "No"! Why?

It takes light time to travel the ten kilometers to the station, and both Isaac and Albert can calculate exactly how long it will take because they know the speed of light doesn't change. But they also know that, by the time the light reaches Isaac, Albert's train will have moved a little way to the east. This means that the light from the western lightning flash will have to travel slightly farther than the light from the eastern flash before it reaches Albert on the train. So Albert will see the western flash *after* the eastern flash. The implications of this small difference are momentous. An event that seemed simultaneous to Isaac is not simultaneous to Albert. In fact, as Issac experiences the eastern lightning flash in the "now," it still lies in Albert's future. Einstein showed that identical clocks timing the same event can produce different results because they are timing the event from slightly different frames of reference. And both are correct. Don't be tempted by the argument that Isaac must be right because he is not moving. *Both* Isaac and Albert are hurtling through space on a planet whose surface is spinning at about sixteen hundred kilometers per hour (depending on how close you are to the equator), while the Earth is traveling around the sun at twenty thousand kilometers per hour, and the solar system is orbiting the center of the Milky Way at more than eight hundred thousand kilometers per hour.

We don't normally notice these temporal anomalies because we rarely encounter things that move fast enough relative to us to make a difference. But the effects are real. I remember as a teenager seeing an experiment on television in which a Geiger counter measured the rate of radioactive breakdown of a small chunk of uranium. You could hear the regular clicks. Then the uranium was put into a centrifuge and spun very, very fast; now it was traveling much faster than the objects outside the centrifuge. That put it within a different frame of reference from me or observers within the TV studio. It was now moving like Albert, the traveler on Einstein's train, but much, much faster. As the centrifuge accelerated, the clicks from the Geiger counter began to slow down. Measured from within the TV studio's frame of reference, time inside the centrifuge seemed to be slowing down. As a teenage science nerd, I was stunned and enchanted. Today, GPS systems have to take account of these subtle differences, because they reconcile the very different frames of reference of satellites hurtling around the Earth and cars crawling over its surface. The physicists who work at particle accelerators such as CERN's Large Hadron Collider (which is in effect a huge centrifuge) also have to take these effects seriously as they accelerate subatomic particles to near light speeds.

In his general theory of relativity, published in 1915, Einstein showed that gravitational fields can also warp measurements of space and time. That idea is taken up in the 2014 film *Interstellar,* whose hero, Cooper, travels through black holes, the densest entities we know, and returns to the solar system to find his daughter is many decades older than he is.

How do Einstein's arguments affect our understanding of the future? Above all, they mean there is no absolute way of specifying when the past ends and the future begins. The answers will vary depending on where you are and how you are moving. An event that is in my future may be in your past, so our definitions of future and past depend on our frames of reference. They are relative.

Einstein's ideas also affect our understanding of causation because he showed that nothing can travel faster than light. That means that causal effects cannot travel at infinite speed. If I kick the winning goal in the World Cup (the probability is low), news of my triumph will be sent into space at the speed of light. Within a second my fans on the moon will be celebrating, and in a little more than four years they will be celebrating on planets orbiting our nearest star system, Alpha Centauri. But it will take 2.5 million years before my fans in the Andromeda galaxy get the good news. Until then, my triumph will have no impact on them. It's as if the ripples from my triumph were spreading out at the speed of light from the place where I kicked the goal — it takes longer and longer for the ripples to reach places far away from me.

Einstein and his friend the mathematician Hermann Minkowski illustrated this idea graphically with the idea of a light cone. I can affect events in the future only in a region of spacetime that widens at the speed of light as we imagine moving into the future. Similarly, I can be affected by something in the past only if I am within its future cone. The Einstein/Minkowski light cones separate those domains of the block-universe with

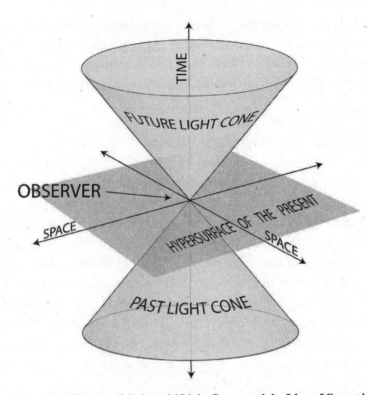

Figure 2.1: Einstein/Minkowski Light Cones and the Idea of Causation
A two-dimensional representation of a four-dimensional "hyperspace."
(K. Aainsqatsi, SVG version of Image: World_line.png, May 7, 2007,
https://commons.wikimedia.org/w/index.php?curid=2210907)

which I can have some causal connection (those inside the two cones) from those to which I can never be connected.

In summary, Einstein showed that if you are asking questions about time and the future, you need to specify the "frame of reference," or perspective, from which those questions are being posed, because different perspectives will yield different answers. The "truth" about the present and future will vary depending on your frame of reference. These arguments illustrate the philosophical position known as *perspectivism*, which philosopher David Danks defines "roughly" as "the idea that scientific theories, models, knowledge, and claims are from a perspective, rather than necessarily expressing objective universal truths."[8] And they show that the future and our ways of understanding the future must be understood perspectively.

The Future for Living Organisms

In our daily lives, Einsteinian relativistic effects make little practical difference because you, your friends, your hometown, and your home planet are all moving through space at more or less the same speed, so you are within more or less the same frame of reference. But frames of reference are not determined just by motion. They vary in other ways, too. Our frames of reference as humans are shaped, above all, by the simple but profound fact that we are alive. Being alive gives all living organisms a distinctive relationship to the universe, time, and the future. It determines what the future means for them and how they try to deal with it.

What does it mean to be alive? That question has generated discussions as complex and difficult as those about time.

There are many definitions of life. NASA, for example, defines life as "a self-sustaining chemical system capable of Darwinian evolution." Here, we need to focus on those features of living organisms that shape their relationships to the future, and two stand out: (1) living organisms are complex so they are subject to decay and breakdown in the future, and (2) whether consciously or not, living organisms act as if they have purposes and goals, so they care about and try to shape their futures. Both ideas — complexity and purposefulness — need a bit of unpacking.

What do we mean by describing a molecule of DNA or a goldfish or my next-door neighbor as complex? Modern physics is full of "fields" that generate the simplest and most fundamental components of the universe, including energies such as gravity and elementary forms of matter such as quarks.[9] These are about as simple as it gets. They are not composed of other things, though they do have properties that can change during fleeting encounters with other forces and entities. Much of the universe consists of simple things and forces that seem to jiggle randomly, backward and forward in time. It is not clear what significance the idea of time can have for such simple entities.

Complex entities are different. They are not better or worse than simple things, and the second law of thermodynamics ensures that complex things are less common and less robust than simple things. But for us humans complex things give the universe most of its beauty, meaning, and significance. We can define complex things as structures composed of diverse components arranged in precise ways that give them distinctive "emergent" properties. Complex things are built so their structures can survive for a while (it could be seconds, it could be trillions of years); if they couldn't survive we would not even notice them.

Atoms and molecules are complex entities; so are stars and starfish; so are crystals and bacteria and the observers in Einstein's experiments along with their clocks and rulers and speed guns. Atoms are made from protons, neutrons, and electrons arranged in precise ways that give them emergent properties such as varying degrees of radioactivity or different ways of reacting with other atoms. Chemists and physicists can measure these properties with great precision. Because the relations between their components can change, their structures and properties can change. So, for complex entities, time does have significance: it means change. And eventually it means breakdown, because sooner or later all complex structures will disintegrate into their component parts in obedience to the second law of thermodynamics, whose diabolical workings were mentioned in the previous chapter. This means that for complex entities the future is the era in which they will eventually break down, so it is full of dramatic tension. How long will they survive? When will they break down? How? Indeed, the story of the whole universe can be told as a drama played out between complex entities and the entropic forces that will eventually take them down.[10]

The second crucial feature of living organisms is that they act as if they had purposes; they show "agency." "Purposeful behavior," writes the geneticist Paul Nurse, "is one of life's defining features."[11] Of course, we use words like *purposefulness* or *agency* metaphorically when discussing living organisms, because the truth is that words like *purpose* or *agency* are really placeholders for phenomena we do not fully understand, like *dark energy* in physics. So in what follows, it must always be understood that, by *purposefulness* or *agency,* we mean behavior that *looks* purposeful.

Complex things that are not alive do not give the appearance of purposefulness. True, they can survive for a time, and that means they can avoid breakdown for a bit. But they survive mechanically, as a result of the physical laws that constructed them. Atoms, for example, are held together by electromagnetic forces, sometimes for billions of years. But heat them enough, perhaps inside a star, and we can predict confidently when they will disintegrate. They won't try to avoid the heat or bargain with the universe.

Living things are different. They will not "go gentle into that dark night," in the words of Dylan Thomas. When threatened, they seem to "rage against the dying of the light." Watch a bacterium tack and weave as it chases food molecules or retreats from danger, and you are watching something very different from an atom, and much less predictable. The behavior of bacteria is less mechanical and more creative and open-ended. It must be, because a bacterium is never in balance with the constantly changing forces and energies swirling around it. Like Huckleberry Finn and Jim, every bacterium seems to deliberately maneuver the fragile raft of its existence through constantly changing flows of energy and matter, always on the lookout for novel solutions to new challenges. Bacteria act as if they really *want* to survive, and they fight for survival with dazzling creativity and ingenuity, which explains why their individual behavior is so hard to predict. Like all living organisms, they are locked in a constant battle against the possibility of entropic breakdown, and that's what makes their relationship to the future so tense, uncertain, and dramatic. While the nonliving meet the future passively, the living seem to meet it actively, with discrimination and purpose. Unlike atoms or asteroids, they seem choosy about the futures they meet.

What is the source of this apparent purposefulness? We do not yet have a complete answer. Many traditional religious and philosophical traditions have seen purposefulness as a quality built into the universe by the beings who created it. But modern science can detect no underlying purpose to the universe as a whole. So the challenge is to explain how a universe without purpose has given birth to entities that act as if they have purpose.

The best explanation available today for the apparent purposefulness of living organisms is that it arose from, and is sustained by, the blind, purpose*less* mechanism that Charles Darwin called *natural selection*. As the philosopher Daniel Dennett puts it, "The process of evolution by natural selection . . . [which] has no foresight at all . . . has gradually built beings with foresight."[12] It does so by equipping living organisms with more and more tricks for surviving and by designing them to use those tricks. Living organisms make copies of themselves, so that if a particular organism is taken down, copies will preserve its structure and the skills that enabled it to survive for a while. That is why all living organisms act as if they have two fundamental purposes: to survive and to repro-duce. The real beauty of natural selection arises from imper-fections in the copying process, because the imperfections create slight variations, some of which may offer new ways of surviving. Over billions of generations, the adaptations that increase the chances of surviving get passed on because only the survivors reproduce. This explains why natural selection has endowed living organisms with such a vast and flexible repertoire of tricks for responding to uncertain futures in dif-ferent environments.

One of the most fundamental of these survival tricks is pur-pose itself. We do not find organisms that don't seem to care what happens. The organisms we see today are here because their ancestors acted as if they really wanted to survive and reproduce — and managed to do so despite the constant sneak attacks of destructive forces from the dark cover of the future. Natural selection over four billion years explains why all living organisms are such creative future thinkers and managers.

In summary, living organisms face the future from a dis-tinctive frame of reference. First, they are complex entities, always in danger of breakdown, which is why their relationship to the future has all the dynamism and uncertainty of A-series time. Second, living organisms seem to act with purpose, as if they care about the future: they seem to actively and creatively seek futures in which they can survive, reproduce, and perhaps even flourish. For living organisms, the future is a realm in which little is certain, much is at stake, and you have to take your chances. But you are not completely powerless. As the ecologist Carl Safina writes, in a wonderful description of the dangerous world of flying fish, "All success — the flying fishes', the birds', everyone's — is temporary, but temporary success is everything."[13]

Anticipating and Managing Futures: General Principles

How can living organisms influence something as insubstantial as the future? There are several basic principles that apply to the future thinking of all living organisms. Some we have already seen, but here we describe them more formally.

The first principle is simple: we have no evidence from the future, so we cannot expect detailed knowledge of possible futures except in rare instances such as the prediction of eclipses, though even then the prediction is based on knowledge from the past. As the historiographer R. G. Collingwood complained, there are no future documents against which we can check our hypotheses about possible futures.[14] My birth certificate tells me when and where I was born, but I do not (yet) have a death certificate telling me when, where, and how I will die. This means that arguments about the future are not tightly constrained by evidence, unlike historical or scientific or legal arguments. No wonder the rules of future thinking are so different.

The second principle is implicit in the first. It is the paradoxical idea that the only evidence about likely futures lies in the past. We are all like the soothsayers in Dante's *Inferno*, with our heads twisted backward. Trying to anticipate the future by studying the past is one example of a research strategy that we use more than we usually admit. I will call it the Nasreddin Hoca (pronounced "Hódja") method, after a story about a famous Turkish sage. One night, Nasreddin Hoca lost his wedding ring in the dark basement of his house. He searched but couldn't find it. Eventually, he went outside and started looking under a streetlamp. A friend asked why he was searching there, and Hoca replied, "Because this is where the light is."

To use the Nasreddin Hoca strategy is to search where the light is. For our purposes, it means looking for evidence about hidden futures in the well-lit world of the past. That's why all living organisms have sensor molecules or organs that help them detect the currents swirling around them in the past and present, because these may shape currents hidden in the future. St. Augustine understood this: "When . . . people speak of knowing the future, what is seen is not events which do not yet exist . . . but perhaps their causes or signs which already exist . . . and that is the basis on which the future can be conceived in the mind and made the subject of prediction."[15]

The third general principle of future thinking is that our ideas about the future can shape the future. However mighty the river of time may be, even bacteria have some control over the fragile rafts of their lives, and that means that what they do will shape the future. Today, most scientists are convinced that continued burning of fossil fuels will transform global climates in dangerous ways. That hunch about likely futures is affecting actions today, which will shape the history of climate change in coming decades. Past thinking (aka history) can change how we think about the past but, as far as we know, it cannot actually change the past. Future thinking in contrast can shape its own subject matter: the future.

The fourth principle is the most fundamental and the most complex: despite having no evidence from the future, we *can* find promising hints about the future in the past. There are two main ways of doing this. (1) We can ask other purposeful beings what they intend to do, and (2) we can study past currents and trends and cautiously project them into the future.

The simple way of getting clues about likely futures is to ask other purposeful beings what they intend to do. Language allows humans to use this strategy with exceptional virtuosity. We constantly ask others (including supernatural beings) about their intentions. "Members of the jury, do you find this person guilty or not guilty?" "Apollo, could you blast my enemies off the face of the Earth?" But the strategy of learning about the future by consulting with other beings is useful only in a few situations: when we think the future depends on someone's decisions, and when we believe we can communicate with and influence that being. This strategy will loom large in chapter 6, which discusses ancient practices of divination.

The most important way of using the past as a guide to the future is *trend hunting*. This is like studying the currents in which your boat is drifting in order to decide where they may carry you and if you can change course.

Trend hunting is practiced by all living organisms, and even for humans it is the strategy of first resort. Our minds look for and calculate likely trends all the time, mostly subconsciously. Trend hunting does not look for certainties. There are none in the future, and if there were, we would not have the time or resources to wait until we knew we had found them. Trend hunting looks, instead, for general patterns that we hope will continue in the future, so it is a bit like betting on the horses. Often it means looking for currents or trends that we think we can steer. The economist Brian Arthur offers a metaphor that Huckleberry Finn might have enjoyed:

> If you think that you're a steamboat and can go up the river, you're kidding yourself. Actually, you're just the captain of a paper boat drifting down the river. If you try to resist, you're not going to get anywhere. On the other hand, if you quietly observe the flow, realizing that you're part of it, realizing that the flow is ever-changing and always leading to new complexities, then every so often you can stick an oar into the river and punt yourself from one eddy to another.[16]

Despite its probabilistic nature, trend hunting can be extremely powerful because some past trends are so regular that we can project them into the future with great confidence. Death, taxes, and the sunrise at dawn are as certain as any facts about the past.

There are four main ways of trend hunting.

The first and most universal method is through direct detection of correlations or trends, such as the falling temperatures that persuade a bear it's time to hibernate. Information about what is going on now sets limits to possible futures, so (good) information is critical to all forms of future thinking. If you've already played the ace of spades this deal, I know no one else can play it. That means as a general rule that the more information we have about trends, the better we can predict. That's why governments collected such detailed statistics about changing infection rates at the height of the COVID-19 pandemic.

The second method of trend hunting is *random dipping*. Like gold prospectors, you dig and hope. You take random samples in the hope of stumbling on promising trends. Modern mathematical techniques such as "Monte Carlo simulations" use random dipping with great sophistication. Natural selection uses it all the time as it spins its genetic roulette wheel, trying out one random genetic variation after another until it comes across those that work. Political pollsters engage in random dipping when they interview random voters to get hints about likely election results.

All species capable of sharing information (and even plants can do that to some degree) deploy a third method of trend hunting, based on shared knowledge of what is going on now and what is likely to happen in the near future. Fourth, humans (and possibly *only* humans) systematically study the *causes* of trends. We study why trends are as they are, because if we can figure that out, we will have a much better idea of how a trend may play out in the future. Newton's laws of motion explained

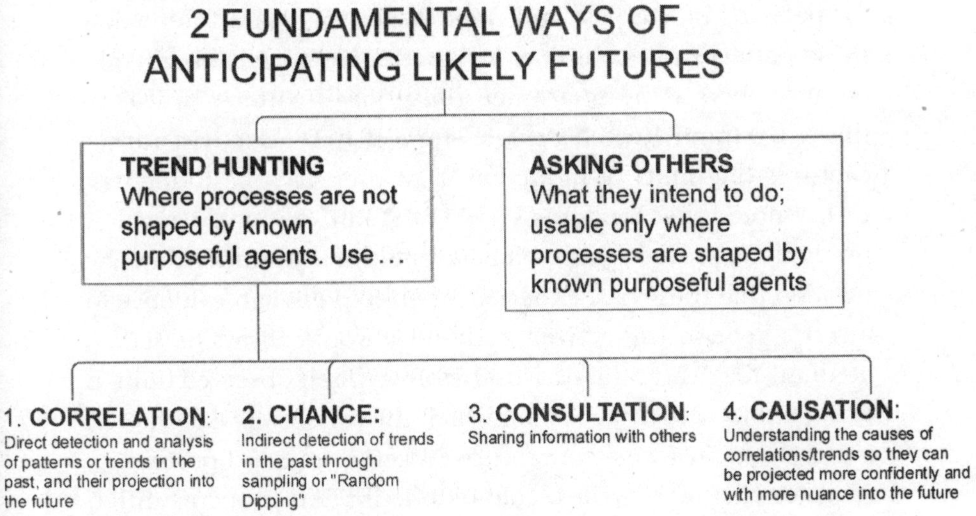

2 FUNDAMENTAL WAYS OF ANTICIPATING LIKELY FUTURES

TREND HUNTING	ASKING OTHERS
Where processes are not shaped by known purposeful agents. Use ...	What they intend to do; usable only where processes are shaped by known purposeful agents

1. CORRELATION:	2. CHANCE:	3. CONSULTATION:	4. CAUSATION:
Direct detection and analysis of patterns or trends in the past, and their projection into the future	Indirect detection of trends in the past through sampling or "Random Dipping"	Sharing information with others	Understanding the causes of correlations/trends so they can be projected more confidently and with more nuance into the future

Figure 2.2: How to Use the Past to Anticipate Likely Futures

why cannon balls, apples, and planets move in precisely measurable ways; and that causal understanding made it possible to project those trajectories into the future with increased precision. In chapter 7, we will see that improved understanding of causation explains much of the predictive power of modern science.

Why should we trust past trends to offer guidance about likely futures when we know things can change in a split second? The logic behind trend hunting is what philosophers call "inductive logic," or simply "induction." This is different from the "deductive logic" behind, say, Euclid's mathematical theorems. Careful deduction promises perfectly true results if its axioms are correct. Inductive logic does not promise perfect knowledge, but in practice it is more useful than deductive logic because so few axioms are certainly true. Inductive logic looks for patterns in the past and then makes the leap of faith that those patterns will continue. It does not always get things right because there is no guarantee that today's patterns will continue. Bertrand Russell's tragic story of the inductivist turkey captures the limits of induction. Day after day, the inductivist turkey noted that food was served at 9 a.m., and every day her predictions proved correct. Then, one fine December day, she decided that these past experiences offered enough evidence to assert a general (inductive) truth: food would always be served at 9 a.m. Tragically, this was just before a widely observed human ritual known as Christmas, for which the turkey was slaughtered and roasted, just when the future looked bright and promising. As Alan Chalmers writes, cold-bloodedly, "An inductive inference with true premises has led to a false conclusion."[17]

Despite the limitations of induction, we have to use it because the world is not as neat or logical or predictable as Euclid's mathematics. The inductivist turkey was actually on the right track, because inductive arguments get it right a lot of the time. The logic behind induction is a principle described well by David Hume: *that principle that instances, of which we have had no experience, must resemble those, of which we have had experience, and that the course of nature continues always uniformly the same.*[18] This is Nasreddin Hoca's principle: the things we can see may offer clues to what is hidden. Time's visible face may tell us something about its hidden face. But we can't know for sure. Hume's principle of uniformity is not really a philosophical or scientific law, but a powerful hunch, or, to put it more honestly, a "leap of faith." As Hume puts it, "We are determined by custom alone to suppose the Future conformable to the past."[19] I have seen the sun rise at dawn so many times in the past that I fully expect it to keep rising at dawn in the future. I have no guarantees, but this hunch is good enough to bet on in a world of imperfect knowledge. Indeed, I can probably treat my prediction as a simple truth. In its 2010 guidelines to authors, the 2010 Intergovernmental Panel on Climate Change report suggests that "in some cases, it may be appropriate to describe findings for which the evidence and understanding are overwhelming as statements of fact without using uncertainty qualifiers" (see table 2.1 on page 62). There are many hunches that have worked so well in the past that we treat them as simple facts, as in the famous Buddhist meditations on death, which are full of confident predictions:

1. Death is inevitable.
2. Our life span is decreasing continuously.
3. Death will come, whether or not we are prepared for it.
4. Human life expectancy is uncertain.
5. There are many causes of death.
6. The human body is fragile and vulnerable.[20]

But our use of induction is more than a matter of custom or blind trust. There is a deeper reason why induction works so well so often. Though nothing is predetermined in detail, our universe is not chaotic. There are general rules, and these ensure the existence of regularities and trends, such as those we observe in astronomy, that can reasonably be projected into the future. Many of those rules are probabilistic. They don't predetermine everything that will happen, but they steer events, sometimes with a looser rein, sometimes with a tighter rein. They set limits to what can happen, and that is why some futures really are more likely than others. For example, one of the universe's most fundamental rules says that the force of gravity always pulls things together. This rule explains the multibillion-year-long trend of star creation in our universe, as gravity crushed protons together in the hearts of stars. But the rule does not determine where and when each new star will appear. The existence of fundamental laws of change explains why the hunches of induction, though never guaranteed, so often turn out to be right.

Once identified, trends can be used to generate maps or models of possible or likely futures. In limited ways, all organisms model their futures and prepare for the futures that look most likely, though we humans do this with dazzling virtuosity. We construct many different models of possible futures in our minds because our sensory apparatus offers too little information for certainty. Our eyes, for example, can detect only a tiny part of the electromagnetic spectrum. But, as we will see in chapter 4, our future anticipation system, installed by natural selection, will interpolate, filling in the gaps by making guesses based on past experience. A momentary glimpse of a semitrailer hurtling toward me is enough information to model unpleasant futures in which I am crushed like a bug; but I can also model better futures so I get out of the way.

Optical illusions show us how the mind interpolates to build models even where information is missing. The *Necker cube* consists of just twelve straight lines.[21] But stare at it for a while, and you can watch your mind using that meager information to model what is in front of you. You are probably seeing a three-dimensional cube. But the optical illusion is designed to be ambiguous, giving us a chance to watch our mind struggling to build alternative models. In this case the mind cannot decide whether the cube's front face is down and to the left or up and to the right. Stare at it for five seconds or so, and it will flip orientation. Children at play build such toy worlds all the time; so do chess players calculating variations; so do artists or scientists exploring alternative hypotheses about how the world might be. Model building is fundamental to modern science. And the building of many model futures using information about past trends is fundamental to all future thinking. No models are perfect, but, as the English statistician George Box puts it, "All models are wrong, some models are useful."[22]

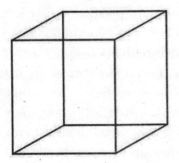

**Figure 2.3: Watching Our Minds
Building Models: The Necker Cube**

When trend hunting, it is also important to distinguish between different types of trends. Trends have different shapes. They can travel in straight lines, or they can be exponential, wavelike, or too erratic to encourage confident predictions. Of course, it is the regular trends that are most helpful in future thinking. Some domains of reality, such as astronomy, yield many reliable mechanical trends that can be used to make powerful and sometimes quite detailed predictions. Astronomers can tell you when you will see the next lunar eclipse at your location, and they can even tell you how long it will last. Other domains, such as politics, yield few regular trends because the actions of humans, like those of all living organisms, are really hard to predict. Some domains of reality offer no basis for confident prediction. Don't ask me to predict interest rates in ten years' time. And much of the future is completely dark. We also need to distinguish between real and imagined trends, between what statisticians call *signals* and *noise*.[23] Is that a bear lurking in the woods at dusk, or a bush shaking in the wind?

All our future thinking culminates in the decisive, dramatic, and mysterious moment when multiple possible futures turn into a single present and we must predict and act. Our future thinking has prepared us for this moment by narrowing down the possibilities, but rarely does it narrow them to just one possible future. So how specific should my predictions be? Like all gamblers, we face a delicate choice. Should I gamble everything on predicting a single winner (whether a horse or a company) in the hope of a huge win; or should I spread my bets across several horses or companies, to increase the likelihood of winning despite reducing the likely gains? As every fairground fortune-teller knows, when making predictions, you must avoid being overspecific (when your prediction will almost certainly prove wrong) or overgeneral (when your prediction will lack interest). "You will meet a tall, dark, rich stranger tomorrow and marry in a week" is too precise to be accurate. "You will meet a stranger" is probably accurate, but too general to be interesting. "*Que será será*" is as safe a prediction as you can make but . . . who cares?[24] Nicholas Rescher describes this as "the vexatious general principle that, other things being equal, the more informative a forecast is, the less secure it is, and conversely, the less informative, the more secure it is."[25] Finding the sweet spot between generality and precision is perhaps the most subtle challenge faced by all types of future thinking.

Remarkably, it is a task that is handled pretty well by all living organisms because of the many refinements built into their future-thinking machinery by natural selection. Without those skills, life could not have survived on Earth for almost four billion years.

The Geography of Imagined Futures: Future Cones

Our ideas about possible futures are foggy because, unlike our ideas about the past, they are not constrained by detailed evidence, by dates, names, and events. So we can think of our attempts to glimpse likely futures as cloudy imagined geographies. Some of these geographies of the future may be as fanciful as medieval maps of distant lands, full of mythical monsters. But some may turn out to be surprisingly accurate. How we imagine the geography of the future is important because, as our third general principle tells us, imagined geographies shape our actions, and our actions today will shape the futures we meet tomorrow.

To get a feeling for these imagined geographies, we will return to the idea of future cones. The future cones that follow are descended from the Einstein/Minkowski light cones we saw earlier in this chapter. It is important to stress that they do not try to describe the future. What they describe is the types of landscapes we think we may find when we reach the future.

The three diagrams that follow highlight some of the most important features of imagined future geographies. They follow

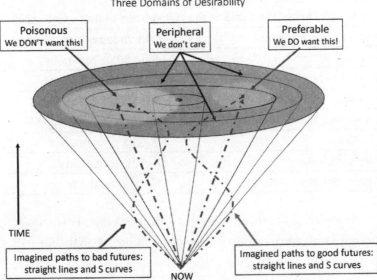

WHAT IMAGINED FUTURES DO WE WANT?
Three Domains of Desirability

Poisonous
We DON'T want this!

Peripheral
We don't care

Preferable
We DO want this!

TIME

Imagined paths to bad futures:
straight lines and S curves

NOW

Imagined paths to good futures:
straight lines and S curves

Figure 2.4: Future Cone 1: Domains of Preference

the futurists' whimsical alliterative convention of trying (for no good reason that I know of) to use adjectives beginning with a *p* (for *probability?*) to describe different domains in the future. Note that the exact positions of different domains are arbitrary and should not be overinterpreted.

As purposeful beings, the first question we must ask about the future is: what futures would we like? Our first future cone tries to capture our intuitive sense that the future contains Badlands and Goodlands and many landscapes in between. Normally, we head for the Goodlands, the Utopias, and try to avoid the Badlands.

Our second question is: which futures are most likely? This is where we look to past trends for guidance. But past trends vary in their regularity and the guidance they offer. Scales of likelihood are widely used in fields such as weather forecasting and estimates of climate change. For example, in its 2010 advice to authors, the International Panel on Climate Change (IPCC) proposed the following seven-step scale:

Table 2.1. The IPCC's Scale of the "Likelihood of Outcomes" (2010)

Term	Likelihood of the outcome
Virtually certain	99–100% probability
Very likely	90–100% probability
Likely	66–100% probability
About as likely as not	33–66% probability
Unlikely	0–33% probability
Very unlikely	0–10% probability
Exceptionally unlikely	0–1% probability

Our second future cone uses a simplified scale with just four domains of predictability. The percentage probabilities should not be taken too seriously; they are just indicative.

In the darker outer domains of Future Cone 2, we find few or no helpful trends and so little guidance to possible futures

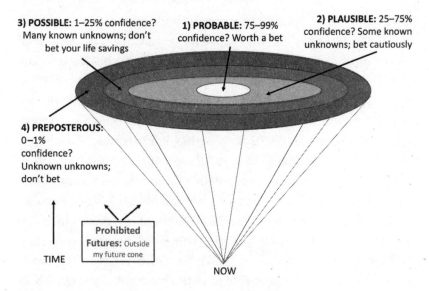

WHAT IMAGINED FUTURES ARE MOST PREDICTABLE?
Four domains of predictability

3) POSSIBLE: 1–25% confidence? Many known unknowns; don't bet your life savings

1) PROBABLE: 75–99% confidence? Worth a bet

2) PLAUSIBLE: 25–75% confidence? Some known unknowns; bet cautiously

4) PREPOSTEROUS: 0–1% confidence? Unknown unknowns; don't bet

TIME

Prohibited Futures: Outside my future cone

NOW

Figure 2.5: Future Cone 2: Domains of Predictability

that prediction generally seems "preposterous." Will a super-nova wipe out the solar system tomorrow? We don't really know. Philosopher Toby Ord speculates that there is less than one chance in fifty million of it happening in the next hundred years.[26] But even he knows that's a wild guess. Step inward and we reach the "possible" domain, where things are still pretty irregular — there are no trends or patterns reliable enough to encourage confident forecasts. These futures could turn up but we shouldn't bet too much on them. Will fusion technology provide abundant nonpolluting energy within thirty years? It's possible, but we can't say much more than that. Move inward one more step and the trends get more reliable, offering "plausible" hints but no certainty. Here you find many probabilistic processes for which we can predict aggregate outcomes but not individual events: I'll bet on how long it takes half a lump of uranium to break down, but not on when this or that particular atom will break down (that bet lies in the "preposterous" domain). In the "plausible" domain we may bet, but with caution. Will the favorite win the Melbourne Cup? The actions of purposeful organisms such as humans and racehorses straddle the "possible" and "plausible" domains, and that makes political forecasting peculiarly interesting, but very tricky. Will the world's political leaders commit seriously to reducing carbon emissions to zero by 2050? It's possible, but how much would you bet on it? In the center of this future cone we find "probable" events. Here, we may bet with some confidence even on the outcome of single events because we are dealing with regular, mechanical, lawlike processes. This is the domain in which the sun rises every morning, the government demands taxes, and, sooner or later, entropy will take us down. This is a domain chemists know well: ignite some 2,4,6-trinitrotoluene (aka TNT) and you will get an explosion. This is where we may think of our forecasts as "moral" certainties.

We will see in chapter 7 that modern science and modern ways of thinking about the future have shifted some processes from less predictable to more predictable domains. Astronomers can now track the motion of most asteroids near the Earth, which has moved predictions about asteroids crashing into the Earth from the "preposterous" to the "plausible" domain. Many medical predictions, including predictions of death rates from particular diseases, have moved in a similar direction, and so have predictions about population growth and the global climate system.

In reality, degrees of predictability vary along a scale with infinite gradations, so assessments of likely futures can be much more nuanced than this four-part scale suggests. The American statistician Nate Silver, who has built predictive systems for politics and sports, notes that specialists in US elections distinguish

WHAT IMAGINED FUTURES GENERATE THE MOST ANXIETY OR HOPE?

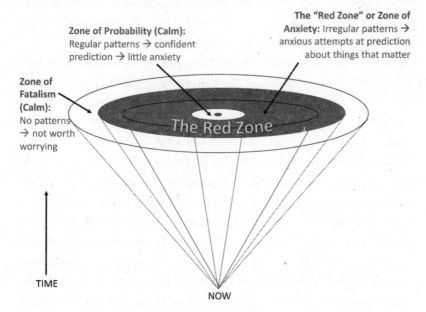

The "Red Zone" or Zone of Anxiety: Irregular patterns → anxious attempts at prediction about things that matter

Zone of Probability (Calm): Regular patterns → confident prediction → little anxiety

Zone of Fatalism (Calm): No patterns → not worth worrying

The Red Zone

TIME

NOW

Figure 2.6: Future Cone 3: Domains of Anxiety

between the predictability of different types of elections: "Polls of House races are less accurate than polls of Senate races, which are in turn less accurate than polls of presidential races. Polls of primaries, also, are considerably less accurate than general election polls."[27] Our four distinct domains of predictability offer a helpful way of thinking about variations in the regularity of trends and the predictability of different types of processes, but they do not pretend to offer precision or nuance.

Finally, Future Cone 3 links preferences and probabilities. As a brainy, mammalian species, we humans have strong desires, fears, and other emotions, and they steer much of our future thinking, sometimes overruling what our conscious minds tell us to do. That is why we often confuse preferred and probable futures. The economist Kenneth Arrow once told a military officer that the weather predictions he was using were statistically random and therefore worthless. He was told that "the Commanding General is well aware that the forecasts are no good; however, he needs them for planning purposes."[28]

Much of our future thinking arises from strong emotion. We invest little emotional or intellectual energy in thinking about futures we cannot predict, those at Future Cone 2's "preposterous" outer edges. Nor do we worry much about futures that are quite predictable because they lie in the "probable" domain. Those domains invite the stoicism shown by some prisoners on death row. Strong emotions are engaged most powerfully in the "possible" and "plausible" domains, where there is some prospect of predicting and shaping the futures we care about. This is the Red Zone. It drives powerful emotions and urgent attempts to predict and manage possible futures. These are the domains that send people to astrologers and soothsayers and encourage modern CEOs to pay huge fees to economic forecasters. Confident advice from expert forecasters can reduce the painful uncertainty of not knowing, but it can also provide a scapegoat when forecasts go terribly wrong.[29]

The Red Zone, where urgency and predictability meet, also shapes the relationship of other organisms to the future. And that is the subject of the next chapter.

PART II

Managing Futures

How Bacteria, Plants, and Animals Do It

CHAPTER 3

How Cells Manage the Future

The most surprising lesson we have learned from simulating complex physical systems on computers is that complex behavior need not have complex roots. Indeed, tremendously interesting and beguiling complex behavior can emerge from collections of extremely simple components.

— CHRISTOPHER LANGTON,
SANTA FE INSTITUTE, 1989[1]

Compared to an *E. coli* bacterium, you and the nearest Venus flytrap are what Dubai's Burj Khalifa is to an ant on the building's front steps. Cells are tiny, tiny things, most of them far too small to be seen with the naked eye. And yet they, too, seem to twist and turn in search of better futures, and they, too, must seek the elusive balance point between precision and generality as they look for likely futures. Despite their size, their future thinking must be pretty skillful because otherwise life could not have survived for so long. How do you pack so much future thinking into such a tiny space?

Understanding the future thinking of cells is fundamental because all life consists of cells. You and I are vast assemblages of trillions of tiny cells, each of which must cope with its own future pretty well if we are to survive. So cellular future thinking is the foundation for all future thinking. Of course, this level of future thinking is not conscious. It is based on biochemical and neurological mechanisms that do not seem to require anything as elaborate as consciousness. Strictly speaking, talk of the future "thinking" of cells is metaphorical. But I will use the metaphor nevertheless, because the way cells cope with the future seems so purposeful, determined, and, frankly, clever.

All living organisms go about the task of managing futures after (metaphorically) asking three questions: What futures do I want? What futures look most likely? How can I steer toward the futures I want?

In step one, organisms decide what they are hoping for. What is their Utopia? I have kidnapped the word *Utopia* to describe the purposes and hopes that drive all future thinking. The word was first used by Sir Thomas More in a book published in 1516 describing a fictional island society off South America. More's Utopia was intended as a model of a good society, and the word has come to be used for all imagined ideal worlds. Metaphorically speaking, all living organisms have their Utopias. These are the preferable regions of Future Cone 1. In these futures, organisms are safe, well-fed, comfortable, free from excessive stress, and able to survive and reproduce. Even the

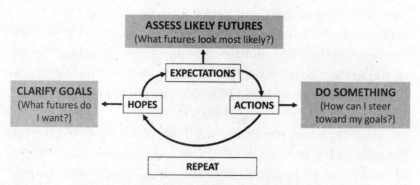

Figure 3.1: The Basic Future Management Kit: Three Universal Steps

simplest organisms can distinguish between a good future and a bad future. Utopias and their opposites, dystopias, give direction and urgency to the future thinking of all living organisms.

In step two, creatures go trend hunting. They seek information about trends that can help them predict the more likely futures. In particular, they study the more regular trends that lie in the "probable" or "plausible" domains of Future Cone 2 because these offer the most guidance. Having identified significant trends and assessed their strength, they will (metaphorically, of course) use inductive logic to project those trends into an imagined future.[2]

Simple organisms like bacteria use general algorithms built into their genomes to assess trends. For example, *E. coli* have algorithms that say it is wasteful to produce the enzymes for processing lactose when there isn't much lactose about. These rules have been installed in the organism's genome over millions of generations by natural selection and have persisted because individuals that inherited this algorithm were more likely to survive and reproduce. But in order to know when to apply the rules, bacteria also need knowledge of what is going on right now. Are lactose levels rising or falling? Identifying trends requires sensors. But it also requires some form of memory so you can compare the situation now with the situation a moment ago. Indeed, it may be that memory exists primarily to enable future thinking. Recent neurological studies have shown that in organisms with nervous systems memory and future thinking are handled by the same parts of the brain, which may explain why people who lose the ability to remember vividly also lose the ability to imagine alternative futures.[3] As Joseph LeDoux puts it, with the elaborate precision of a neurobiologist, "Memory is first and foremost a cellular function that facilitates survival by enabling the past to inform present or future cellular function." Kant guessed at this profound truth more than two centuries ago: "Recalling the past (remembering) occurs only with the intention of making it possible to foresee the future."[4]

Step three is when creatures place their bets, often under the urgent promptings of anxieties from the Red Zone of Future Cone 3. They act. They intervene in the world, dipping their oars into the currents all around them and trying to steer toward their Utopias. If you're a bacterium, you may simply swim off in a new direction because you can't see any food in front of you. As Brian Arthur puts it, managing futures is a bit like the martial arts: "The idea is to observe, to act courageously, and to pick your timing extremely well."[5] Courage is needed at the third step, the courage to act decisively without any certainty about outcomes and with the knowledge that all bets can lose. "Do not be troubled," as Krishna told Arjuna, "but strike!"

Then the cycle repeats. But now you have new information, so you can tweak your plans. You may even adjust your goals if you decide that the original goal was unachievable or too costly, or that you can do better. This constant reassessment of the likelihood of different possible futures is known to statisticians as "Bayesian" analysis, a topic we will return to briefly in chapter 7. The jargon may sound intimidating, but the basic idea is simplicity itself. You begin with an initial rough (or even random) estimate of the probability of something happening. Bayesian statisticians call this your *prior*. Then, as new information comes in, you adjust your prior, tweak your bets, and so on and so on. All living organisms, from amoebas to Venus flytraps, are pretty good Bayesian statisticians.[6]

In the real world the three steps overlap. But thinking about them separately may clarify what is going on as living organisms try to survive in a very uncertain world.

The future management kit of all organisms makes up a significant part of the larger set of skills that we call cognition. "Biological cognition," writes the cognitive biologist Pamela Lyon, "is the complex of sensory and other information-processing mechanisms an organism has for becoming familiar with, valuing, and [interacting with] its environment in order to meet existential goals, the most basic of which are survival, [growth or thriving], and reproduction."[7] The cognitive tool kit of even the simplest organisms includes the ability to anticipate likely future events, to sense and evaluate what is going on in the local environment, to remember and learn, and even some ability to share information with other members of the same species.

As biologists are slowly realizing, all living organisms do cognition.[8] Cognition consists of the learning skills that allow living organisms to react creatively to the ever-changing threats to their survival by preparing skillfully for likely futures. What Daniel Dennett says of minds is true of all living organisms: "A mind is fundamentally an anticipator, an expectation-generator."[9]

We see future management at its simplest in viruses and single-celled organisms. To give some idea of how this works, the rest of this chapter will describe how single-celled organisms manage their futures, using biochemical methods and gadgets similar to those at work in every cell of our own bodies.

The Microbial World

The idea that there are living creatures too small to be seen by the naked eye would have seemed fantastic until a few centuries ago. But today, individual cells are regarded as the smallest things that can be said to be alive and the fundamental building blocks of all life. Cells are as fundamental for biologists as atoms are for chemists.

The first "natural philosopher" to observe cells through a microscope was the English scientist Robert Hooke. In 1665, he saw individual cells in a slice of cork and called them *cella*, the

Latin word for small rooms or cubicles, because he saw that each cell has its own wall, or membrane, dividing it from the rest of the world. The first observer to realize that individual cells might be alive was a Dutch lens grinder named Antoni van Leeuwenhoek. Leeuwenhoek discovered an entire, previously unsuspected world of organisms consisting of just one cell, organisms so small that a million could live together happily in a water drop. He called them "animalcules."[10]

Discovering that large organisms like us share this planet with a miniature world of previously unknown microorganisms was surely as profound as discovering life on other worlds (a discovery that could happen in the next few decades). Not until 1839, though, was it recognized that cells are the building blocks of all forms of life. That's when Matthias Schleiden and Theodor Schwann declared that "all organisms are composed of essentially like parts, namely of cells." In 1858, Rudolf Virchow added the final touch to the cell theory of life by pointing out that every cell could be regarded as a separate living being. A cell's membrane, just two molecules thick, creates a border between the inner, living world and the outer world. But it also enables contact with the outer world and the exchange of energy, nutrients, information, and waste.[11]

It might seem that organisms consisting of just one cell must be simple. But we now know that each may consist of billions of atoms, and thousands of different types of molecules. All these components are meticulously designed, and interact with exquisite precision in chemical dances whose precise choreography we do not fully understand. Each component may seem simple, but as the emerging field of complexity studies has shown, extraordinary complexity can emerge from the activities of simple components interacting through multiple feedback loops.[12]

We know today that single-celled organisms can make skillful and sophisticated bets on their futures. They can learn from their mistakes, they can remember what happened a moment ago, and they can calculate odds, Bayesian-style. They can even create internal molecular models of external conditions such as temperature and oxygen levels and use those models to decide on appropriate action.[13] Of course, if you have just one cell, you can't really think. The brains that do our thinking are built from billions of individual cells, most of them larger than a single bacterium, so brains are not an option for bacteria. Instead, bacteria must manage their futures using networks of biochemical reactions that can, in effect, compute what is needed, what is probably about to happen, and what needs to be done right now.

Minimal versions of the bacterial future-management tool kit have probably been around since life first appeared on Earth almost four billion years ago. Comparisons of the genomes of many different microbial species have shown the existence of future-management tool kits so widespread that they probably existed inside LUCA, the last universal common ancestor, the hypothetical ancestor of all organisms alive today.[14] LUCA existed almost four billion years ago, yet it contained sensors and it could act to protect itself. And it contained a computational network of molecular switches that could make purposeful choices such as "if food is detected (A), move toward it (B), but only if there's a lot of food . . . ; if there is hardly any food (–A), then don't waste effort on moving (–B)." Like all complex systems, that of LUCA almost certainly contained positive feedback loops to drive active, dynamic processes; negative feedback loops to dampen those processes and maintain stability; and links between these systems that could create more nuanced reactions in response to feedback from other parts of the system. In other words, LUCA contained the basic logic circuits needed to perform the calculations all living organisms make as they face uncertain futures.

The star of this chapter is the *E. coli* bacterium, gazillions of which are alive today, several million of them in your gut. They love the human gut.

In recent decades, biologists have learned a lot about *E. coli,* and the genomes of many different variants have now been decoded. Indeed, *E. coli* may have been studied as closely as any other organism apart from ourselves.[15] This is partly because we have learned how to use modified *E. coli* cells as biological factories to produce substances such as insulin. But it is also because when they go rogue, *E. coli* can really hurt us.

The proper biological name of *E. coli* is *Escherichia coli,* after Theodor Escherich, the Austrian biochemist who first identified it. The name refers not to a single species of bacteria, but to several related species that have evolved along different pathways over the last one hundred million years or so.[16] Bacteria form one of the three superkingdoms, or domains, into which biologists group all living organisms: Bacteria, Archaea, and Eukaryotes. Bacteria and Archaea are single-celled and are classified as prokaryotes. With just one cell per organism, prokaryotes have to be generalists. Their single cells have to do everything necessary for the cell's survival, including preparing for the future.

Each rodlike *E. coli* cell is just a few millionths of a meter — a few microns — long. Thirty or forty placed end to end would just about add up to the width of a human hair (about eighty microns).[17] Nevertheless, each cell contains as many as one hundred thousand billion atoms and a lot of interesting biological stuff.

How Does *E. coli* Cope with Uncertain Futures?

To understand how *E. coli* cells manage their future, it would be good to have a guide who can shrink us to the size of a protein molecule and lead us into the strange, squishy world of a bacterium's cytoplasm. That will be unnerving, but it's worth following our guide closely because we'll get to see the basic machinery of all future management, machinery very similar to what you would find in each cell of your own body. It's a strange world. We'll find ourselves in wobbling, muddy country that vibrates with the random disordered energy of heat, while more ordered electromagnetic force fields keep tugging and pushing at us. Around us, we'll see a crowd of molecules that seem to be engaged in a massive mud-wrestling tournament. It's a complex and sometimes violent world, but it's also a world of remarkable collaboration and teamwork.

Let's imagine we've settled our nerves and our guide is ready to lead us. First, we'll head for the cell's genome, the place that stores the information needed to build the four thousand or so different types of molecules that compose an *E. coli* cell and enable it to manage its future. To get to the genome, we'll have to wade through the cell's gooey cytoplasm and past a sweaty and assertive crowd of worker molecules, most of them proteins. Eventually we will reach a huge, gnarled ring of DNA (deoxyribonucleic acid) that floats freely inside the cell. As we arrive, we'll feel a bit like astronauts landing on a beaten-up spiral space station.

Close up, the DNA ring looks like a ramshackle, twisting spiral staircase that loops back on itself. The two sides of the molecular staircase are linked every few steps by rungs, each with two halves made from two bases linked loosely by hydrogen bonds. Each base consists of just a few atoms. There are only four types of bases, so if you were to color them red, white, blue, and black, you could imagine seeing those colors — two different colors to each rung — repeating themselves more than four million times, apparently at random, off into the remote distance. But you'd be wrong about the randomness. As geneticists discovered in the

1960s, the pattern of bases is in fact a four-letter code that contains all the information you need to build the different worker molecules that keep every healthy *E. coli* cell functioning.

To read this code you must first split each rung of the DNA molecule in half; that's not hard because the hydrogen bonds between the two halves of each rung are weak. Then you can read the bases attached to either side of the staircase. Every group of three bases codes for a particular amino acid. For example, if you read down one side of the DNA ladder, temporarily ignoring the other side, you might get a sequence of bases that reads GAT (for the bases guanine, adenine, and thymine), which is the DNA code for the amino acid known as aspartate. The next three bases will probably code for another amino acid, and so on, though some bases code for instructions such as "stop reading here." The exact order of the bases really matters because most of the code is telling you how to build proteins, the molecules that do most of the work of managing the cell's future, and proteins are made from long, precisely ordered chains of amino acids.

The billions of bases that make up the code consist of distinct strings of hundreds or thousands of bases that list (in triplets) the sequences of amino acids needed to build the particular molecules that each organism needs. These strings of bases are the cell's *genes,* and the collection of all genes is known as the cell's *genome. E. coli* cells have about three thousand genes. (We humans are not that much fancier. We have about twenty-one thousand to twenty-five thousand genes.) Most genes code for protein molecules, but some code for RNA molecules, which are like DNA but come only in single strands. RNA molecules are immensely important because they can carry information like DNA, but they can also do serious molecular work like proteins. The list of genes varies from species to species because each species depends for its survival on a unique mix of worker molecules.

The first step in all future thinking is to be clear about goals. Goals are stored, in effect, in the cell's DNA. This is not literally true, of course. There is no placard that says: SURVIVE AND REPRODUCE! EATING WILL HELP! But the genome contains the instructions for manufacturing the proteins and other molecules that the cell needs in order to survive in its normal habitats. In effect, the genome stores information about the short-term goals that must be met for the cell to achieve its long-term goals of survival and reproduction. For example, at some point, an *E. coli* cell will have to break down lactose molecules and — *voilà!* — the genome contains instructions for making a protein that can do just that.

So far, our tour may have suggested that DNA decides what goes on in the rest of the cell, a bit like the control deck of the *Starship Enterprise.* But in recent decades, it has become clear that things are not so simple. DNA consists of information, like a cookbook. It can't actually *do* anything. What shapes the cell's behavior at any one moment is the mix of genes that are actually being used in that moment. And that is determined by the activities of worker molecules known as *transcription factors,* which can sense what is going on inside and outside the cell and use that information to "decide" what new molecules need to be manufactured or scrapped. The transcription factors dive into the DNA, unlock the instructions for making the relevant molecules, and start the process of manufacturing them (or shut down the manufacture of proteins that are no longer needed). At any given moment, only some of the genes in an organism's genome will be "expressed." The rest of the genome is switched off and waiting (sometimes forever) to be read and used. Biologists now refer to processes that determine which genes are being used at a particular moment as *epigenetic* processes. They do not alter the genome, but they do affect how and when particular genes are expressed. Epigenetics is the study of nongenetic factors that determine how and when genes get used.

Epigenetic processes are crucial to a cell's future thinking because they tell the cell what's happening right now and what it needs to prepare for.

Hover just outside the ring of DNA and you'll see a lot of epigenetic activity as proteins and RNA molecules swoop in, armed with new information about looming threats or opportunities, use their molecular wrenches and levers to break apart particular rungs of DNA, and either read off the genetic code for that section or block its expression. If a new protein is needed, a special molecular transcription factor cruises alongside the ring of DNA looking for a particular gene. When it finds it, it prizes apart that section of the spiral staircase by separating some of the base-pair rungs. It then calls in a crew of messenger RNA (mRNA) molecules. The RNA molecules read off and store the sequence of base-pair letters in the exposed gene. Then the opened rungs can be locked up again, and the messenger RNA molecules, now carrying an ordered list of bases — the recipe for a new protein — can head off into the sludge of the cytoplasm to dock at a ribosome, a vast blob of proteins and RNA that functions a bit like a 3D printer. The ribosome grabs the messenger RNA, reads the ordered list of amino acids copied from the DNA, then fishes around in the surrounding sludge, grabs hold of the required amino acids as they swim by, and locks them in place in a long chain in exactly the right order to build a particular protein. Ribosomes work fast. It would take a ribosome only a minute to put together a protein with three hundred different amino acids, and there can be several million ribosomes at work in a cell at any given moment, so a cell can turn out a lot of different proteins simultaneously.[18] This complex manufacturing work goes on all the time in all cells of all living organisms, and it creates the constantly changing mix of molecules that a cell needs to cope with looming crises and prepare for likely futures.

How does a cell know what proteins to manufacture or stop

manufacturing? This takes us to the second step in future management: detecting past trends and evaluating the hints they contain about likely futures. Cells go trend hunting.

To detect trends in the external world, cells use special sensor molecules stuck like cocktail sticks through their membranes, so that part is in the outside world and part is inside the cell. Each *E. coli* cell may have as many as ten thousand sensor molecules sticking through its membrane, most at its forward end, where it first encounters new conditions. With these sensors, *E. coli* cells can detect as many as fifty distinct chemicals. By combining information from different sensors, they can gauge trends in a chemical's concentration with great precision. And, as we have seen, the power of trend hunting increases as more information becomes available about evolving trends.

So let's imagine that our guide now leads us through the sludge of the cytoplasm and out to the cell's membrane, where we can watch some of its sensor molecules at work. To get to the outside part of a sensor molecule, we will clamber through molecular tunnels in the membrane, which lead us to the less claustrophobic world just outside the cell. Now we are among the cell's scouts, spies, sniffer dogs, and frontier guards. Sensor molecules are proteins, like most of the cell's worker molecules, so watching them at work will give us an idea of how all proteins work.

Proteins can make up almost half of a cell's volume (if you don't count water molecules, which account for up to 70 percent of a cell's molecules).[19] At any one time, a given *E. coli* cell may contain millions of protein molecules, each made from several thousand atoms. Some may be under construction, some will be hard at work, while others, having done their job, will be broken down and recycled to make new worker molecules.

How do proteins do their work? Each protein consists of hundreds of amino acids that were linked together in a precise order by a ribosome. Each of the hundreds of amino acids in a protein has slightly different chemical and electrical properties, so as the newly made chain is buffeted inside the cell, it soon folds into a shape that may look like a disorganized ball of steel wool but actually has a very specific structure that includes special biochemical pockets, like baseball gloves, to help the protein capture specific molecules. For example, human hemoglobin molecules (among the first proteins to have their structure deciphered) have pockets that can capture and transport oxygen molecules. Proteins can also rearrange the molecules they have captured by splitting, tapping, bending, zapping, or fusing them together. This is why proteins work as enzymes, making possible chemical reactions that would be impossible without them. Finally, when they capture a molecule, proteins change shape, like a sock into which you have inserted your foot. This shape-shifting (known as *allostery*), creates a sort of short-term memory about events and trends that can be passed on because other molecules notice the changes in the posture of the sensor molecule.

So, back to one of the sensor molecules on the surface of our *E. coli* cell. Let's imagine that this one has a pocket designed to capture molecules of the amino acid aspartate, which *E. coli* cells regard as a tasty snack. If our sensor protein finds an aspartate molecule, it will grab it, and that will change the protein's shape. Its new posture will send a message like the posture of someone bearing joyful tidings, and that message will be felt inside the cell, at the protein's other end. The changing shape of the protein creates a sort of memory because the message about the captured aspartate molecule will be remembered for as long as the sensor protein keeps a grip on its captive molecule and holds its new posture. Inside the cell, other molecules will react by changing *their* shape and cruising through the cytoplasm to spread the good news about the captured aspartate molecule. They, too, remember. Like most molecules in prokaryotic cells, these messenger proteins will travel randomly through the cytoplasmic goo, jostled by heat energy, like humans being pushed around inside an overcrowded bus. In this way, millions of messenger proteins spread information about trends and currents in the outside world. "Aspartate levels high, feast probable," they may shout, or "Aspartate levels falling, famine possible." As the biologist Dennis Bray writes, "It is as though each organism builds an image of the world — a description expressed not in words or in pixels but in the language of chemistry."[20]

Step three in all future management is action: intervening in the world to achieve your goals. How is the information picked up by sensor cells assessed and turned into action?

How cells use information to regulate their behavior was first shown by the French researchers François Jacob and Jacques Monod and their graduate student, Jean-Pierre Changeux, in the 1960s. They first showed how the shape-shifting of proteins allows proteins to work both as enzymes (by speeding up or enabling reactions that would otherwise be impossible inside a cell) and as information carriers. But they also showed that the power of proteins is magnified many times over when they work in teams, or networks, which Jacob and Monod called "operons."

One of the first operons they studied regulates how a cell digests lactose.[21] The key players are protein transcription factors that specialize in blocking genes. They have two pockets, or binding sites. One dangles in the cytoplasm looking out for lactose molecules. If it can't find any, the other pocket will bind to parts of the DNA that code for lactose-digesting proteins and block their expression. The number of repressor proteins detecting lactose tells the cell how much lactose is around. If a lot of the repressor proteins have not found any lactose, they will more or less shut down production of lactose-digesting proteins. But if the repressor proteins start catching more lactose molecules, they will change shape, release their grip on the cell's DNA, and allow the expression of its lactose-digesting genes. If lactose levels fall again, the whole process will go into reverse. This is a neat negative-feedback mechanism, a nice way of ensuring that lactose can be digested when plenty is available, without wasting energy or resources when it's not. This is sophisticated future thinking based on probabilistic decisions about what the cell will need in the near future.

In each cell, at any given moment, there may be millions of operons working together in impossibly complex ways. Some are much more elaborate than the one we have just looked at. For instance, they may have several protein switches that must be turned on before the cell starts producing a new protein, so that the new protein will only be produced if conditions A, B, and C are all satisfied. Here we have a sort of "if A and B and C then D" switch. In other cases, there might be an "if A or B or C then D" switch. In this way, chains and networks of proteins can act like logic circuits. And if enough of these switches are linked together, as in a computer, you can do a lot of computing. As the complexity theorist Melanie Mitchell points out, a machine that can link many "and," "or," and "not" switches in the right ways can compute more or less anything that is computable.[22] This is how the simple biomolecular switches of an *E. coli* cell can perform computations of immense subtlety, including probabilistic calculations about possible futures. Because many operons are at work at the same time, cells compute in parallel. That means that even the simplest cells can compute multiple probabilities at any one moment, about food levels, temperatures, internal salinity, whether to move, and so on.

Let's take motion as an example of the actions that can be triggered by these calculations. *E. coli* cells have up to six sleek and powerful propellers that allow them either to move forward or to "tumble" randomly. Like sensor molecules the propellers are threaded through the membranes. They use whiplike tails, or flagella, that hang outside the cell and can spin at several hundred cycles per second.[23] Let's imagine that the inside ends of the propeller molecules start encountering a lot of messenger proteins shouting that there is a buildup of aspartate ahead. The propellers will work in unison and the cell will move forward. But if aspartate levels are falling, some propellers may switch direction, sending the cell into "Tumble!" mode for a moment. Then it will set off in a new direction, chosen by a sort of random dipping, in search of better hunting.

What just happened? We have seen how an organism far too small to be seen by the naked eye can set goals, assess the current situation, and take pretty good decisions about how to face the future. Its long-term goals are built into its genome, in the form of codes for building the molecular gadgets needed to achieve short-term goals such as finding food. Protein sensors keep the cell informed about the ongoing hunt for aspartate and other foods; networks of proteins assess how things are going; the changing mix and shape of those proteins determines what the cell should do. To tumble or not to tumble? The whole sequence has evolved over millions of years as cells that didn't tumble when it would have been smart to do so were less likely to survive than those that did, while the more successful algorithms got built into the species' genome. That's why the cell's future-thinking machinery works pretty well most of the time, and that's why the *E. coli* lineage has survived for hundreds of millions of years.

This is very clever stuff! The next chapter will explore how multicellular organisms go about the task of future management, using new types of machinery that link the activities of billions of cells, each of them at least as clever as an individual *E. coli* cell.

CHAPTER 4

How Plants and Animals Manage the Future

> Plants in the beginning of a drought often increase root growth toward deep soil layers, searching for new water sources. At the same time, the plant stops the growth of shallow roots where the soil is usually the driest. In this way, a plant hedges its bets and focuses on growing to where there's the highest chance of finding water.
>
> — DANIEL CHAMOVITZ, *WHAT A PLANT KNOWS*[1]

Multicellularity, and How It Changes Future Thinking

Macrobes, like you and me, consist of trillions of cells that collaborate to ensure the survival of a single, vast organism. Macrobes have flourished only in the last sixth of Earth's history, from about six hundred million years ago. Multicellularity allowed life to exist in new forms and at new scales.

Each macrobial cell uses much of the same future-management machinery as bacteria. But in addition to managing their own futures, macrobial cells face a new challenge: they have to coordinate their activities with billions or trillions of other cells to manage the future of the superorganisms on which they all depend. How do you get billions of cells to agree on a good collective future, pool information and assess the trends revealed by that information, and then act collectively for the good of the large organism of which they are all members? How do you get billions of cells to agree that now is the right time to invest energy in running fast from a lion or driving your roots deeper into the soil?

These collaborative challenges are so different from those faced by single-celled organisms that they require new biological machinery for collective future management. Building that machinery took hundreds of millions of years, and that may help explain why macrobes appeared so recently in Earth's history. It certainly helps explain why the genomes of macrobial cells are larger than those of bacteria and code for many new kinds of proteins. The simple nematode worm, *Caenorhabditis elegans* (which we will meet again), is only about one millimeter long and has about a thousand cells. But about 90 percent of its nineteen thousand genes are devoted to maintaining good relations between cells.[2] To survive collectively, macrobial cells need to put a lot of effort into communication, negotiation, and collaboration.

Can the elaborate collaborative future-management machinery built by macrobes teach anything new to a species like ours whose individual members are beginning to realize that their individual futures now depend on the success of humanity as a whole?

How Do Macrobial Cells Collaborate So Well?

To understand the future thinking of macrobes, we first need to understand how it is possible for trillions of cells to collaborate so effectively.

With few exceptions, macrobes are built from eukaryotic rather than prokaryotic cells, so they belong to the third biological superkingdom, the Eukaryotes. The first eukaryotic cells evolved almost two billion years ago, probably through the merger of already existing prokaryotic species. (That revolutionary idea, first suggested by the biologist Lynn Margulis, is now accepted by most biologists.) Compared to the peasant huts of prokaryotes, eukaryotic cells are palaces. They can be hundreds of times larger than prokaryotic cells, and they contain many internal divisions, or rooms, with special uses and functions. The most important is the nucleus, a fortified inner sanctum that protects the cell's DNA.

The eukaryotic cells of macrobes cooperate well for two main reasons. First, every cell in a macrobe has exactly the same DNA. That creates a sort of preprogrammed loyalty to the larger organism. Indeed, macrobial cells, like fighter pilots on suicide missions, are sometimes ordered to die for the good of the larger organism, and mostly they obey. The biologist's jargon for this self-sacrifice is *apoptosis*. The digits on your hands formed because, as they were developing in the womb, cells in the spaces between fingers were ordered to die, and they obeyed.[3] Cancer cells disobey such orders, and that shows how dangerous it is for a macrobe when collaboration breaks down.

The second reason macrobial cells cooperate well is that they can specialize, turning into the cellular equivalents of surgeons and plumbers, musicians and fashion designers. Specialization makes each cell dependent on other cells and on the survival of the larger macroorganism. This is a bit like human beings in modern societies. Farmers grow food while nurses care for the sick. To survive, they must cooperate. The farmer will feed the nurse who looks after her children when they are sick, and both farmers and nurses depend on the orderly functioning of the larger society of which they are both members.

Similarly, macrobial cells divvy up tasks such as fighting invasive bacteria, flexing muscles, or . . . thinking about the future. And the cellular division of labor is almost as complex as that of human societies. In animals, red blood cells specialize in transporting oxygen, bone cells maintain the organism's rigidity, muscles do the heavy lifting, skin cells defend the borders, and neurons transmit information and assess its significance. All in all, there are about thirty trillion cells in a human body, divided into about two hundred different types.

Though all the cells of a particular macrobe share the same DNA, their DNA contains instructions for making many different kinds of cells. This is why eukaryotic cells can specialize. In the earliest days of a macrobe's life, all its cells consist of identical stem cells, which have the potential to turn into many different kinds of cells. But after a week or two, the growing ball of stem cells becomes large enough for different cells to find themselves in slightly different environments. Depending on whether it is inside the ball or at its edge, a cell will feel slightly different pressures and sense different chemical concentrations and temperatures. Inside each cell, these small differences trigger different responses from transcription factors, which dive into the DNA and start blocking some genes and expressing others. Epigenetics is at work once more, and the result is that each cell expresses slightly different genes. As differences accumulate, each cell travels farther along its own distinct career path. Whole regions of its DNA get shut down, in some cases permanently, while other regions are activated. Once you've started turning into a muscle cell, you're committed. That's your future decided. New signals can tell you what sort of muscle cell to become, but they can no longer turn you into a neuron or a blood cell. Specialization explains why most macrobial cells use fewer than half the genes in their DNA — only the genes needed for basic housekeeping are used in all cells.[4] Epigenetic processes also ensure that the specializations of parent cells are shared by their offspring. As DNA copies itself, the copying process transfers the pattern of transcription factors locked onto the DNA to the new copies of DNA, so that the child cells express only the genes expressed by their parents. This is why bone cells produce bone cells, neurons produce neurons, and muscle cells produce muscle cells.

The extreme dependence created by specialization explains why each cell listens intently to signals from outside its membrane. Each cell watches its neighbors closely and samples the chemicals, nutrients, energy, and information flowing past its sensor molecules for messages from farther afield. These messages are like public announcements. They may arrive as electric pulses, or in the form of special molecules such as hormones, or simply from the jostling of neighbors.

In short, macrobial cells are eager to work together and share information with other cells and with the organism as a whole. This sort of collaboration is the foundation for the future thinking of all macrobes. It explains why macrobial cells can share common goals, collaborate on assessing likely futures, and work together once they have decided the best course of action. In the next section, we will look at some of the methods plants use to manage their futures. Then, coming closer to our own species, we will look at how animals plan for their futures, and for the first time we really will start discussing future *thinking*.

How Plants Manage Their Futures

The idea of plants managing their futures may seem odd, because it is so easy to think of plants as passive. But that illusion hides a lot of purposeful and sophisticated future management.

Plants differ from animals mainly because they get most of their energy directly from the sun using the complex biochemical reactions known as photosynthesis. Plants lap sunlight as kittens lap milk, but sunlight is all around them, so they don't have to move to get it.

Photosynthesis provides most of the biochemical energy flows on which life depends. It is powered by energetic photons of sunlight and managed by the chloroplasts found inside plant cells. It also requires water and carbon dioxide, but these also get delivered, like light, to the plant's doorstep. Other vital elements, such as nitrogen, phosphorus, and magnesium, can usually be found in the soil by a plant's roots. Because the energy and nutrients they need are all around them, most plants are "sessile" stay-at-homes. Their young may go wandering as spores or seeds. But once settled, plants usually stay in one place for the rest of their lives.

That doesn't mean they can relax. To survive and reproduce, plants have to seek out information and make probabilistic bets, like all living organisms.[5] But plants place their bets using the currency of energy and nutrients rather than hard cash, and their bets are mostly about how to manage their own bodies. If I am a weeping willow, how much energy should I stake on growing taller than my neighbors? Is now a good time to grow more leaves, or should I be making flowers? Should I get ready to ward off an attack of beetles? Whether you are a weeping willow, a snowdrop, or a potato, you will place bets on likely futures as routinely as any human gambler, and your future will depend on the number of bets you win.

Plants manage their futures using the same three steps as all living organisms. They have goals, both large and small; they look for and analyze trends in their environment to figure out what may be coming next; and then they act, they place their bets.

Snowdrops and potatoes have their own Utopias. But in detail, success looks different to different species, and the steps toward success vary too. So each plant has its own microgoals, most of them built into the genome in the form of genes for making the proteins and cells each species needs to survive, flourish, and reproduce in its particular niche. The microgoals are like lists of biochemical tricks and maneuvers that many generations of weeping willows or cacti have found helpful in the past.

Step two means trend hunting. To learn what's going on outside their bodies, the cells on a plant's surface have sensor proteins that sniff out changes in the molecules, energies, scents, and even sounds that surround them. As connoisseurs of sunlight, plants are particularly good at discriminating between light frequencies. Arabidopsis, which is related to the mustard plant, and widely used in botanical experiments, has at least eleven different types of light detectors: "Some tell a plant when to germinate, some tell it when to bend to the light, some tell it when to flower, and some let it know when it's night time."[6]

Once detected, information about evolving trends has to be sent to other cells, some of them a long way away. Communicating with neighbors is simple. Some cells can pass proteins, including transcription factors, directly to neighbors through their membranes.[7] To transport water, nutrients, and information-carrying molecules from roots to leaves, vascular plants use woody channels known as the *xylem*. Liquids are drawn upward partly by pressure from below as evaporation lowers pressure in the upper parts of a plant. Vascular plants also carry sap, which distributes information-bearing hormones and the energy-rich by-products of photosynthesis to all parts of the plant. The sap is carried downward from leaves in special tissues known as the *phloem*. In trees, the phloem lies just under the bark.

Plants can also transmit information electrically. In the 1990s, a team of researchers led by Dianna Bowles showed that the wounded leaves of tomato plants could effectively phone other leaves by sending them electric signals. The other leaves responded by manufacturing protective proteins in case they, too, were attacked. Since then, Swiss researchers have shown, using Arabidopsis plants, that the electrical pulses for these botanical phone calls are generated by *chemiosmosis,* a mechanism present in most kinds of cells. Special pumps let cells control the concentration of potassium and calcium ions (charged atoms) on either side of their membranes. This creates a slight electric potential across cell membranes that can be used, among other things, to fire an electrical pulse.[8] We will look at these cellular batteries in more detail when we meet neurons later in this chapter.

Using signal molecules such as hormones, and electric messages, plant cells can share a lot of information about what is going on and what needs to be done. Leaves and roots can sense chemicals in the air and the soil and alert other cells to their presence, and plants know when they are being touched (watch a Venus flytrap close its spiky jaws on a small frog). Recent research suggests plants may even be able to hear sounds such as the flow of a nearby stream. Plants can also share information with other plants. For example, they can sense the changing intensity and composition of the mist of proteins and chemicals known as *pheromones* emitted by neighboring plants. Plants attacked by beetles may emit beetle-repelling proteins and chemicals. Nearby plants will sniff the chemicals, assess the threat they pose, and respond, perhaps by producing their own protective toxins. For humans, who have more efficient ways of communicating, pheromones are bit players. But for plants, they provide a sort of chemical language. They may not let you debate the meaning of life or the philosophy of time, but they can help you share information about likely futures. More recently, Suzanne Simard has shown that trees also use their roots to share information and nutrients through huge networks of fungi in what has come to be known as the "Wood Wide Web."[9]

Plants that start manufacturing toxins after sniffing the

pheromones emitted by distressed neighbors have acquired information about likely futures: a threat that was once just possible has now become probable. And they have placed bets on the basis of that information, paying for those bets in the energy needed to manufacture new chemicals.

Assessing likely futures means looking for trends. And recognizing trends requires some form of memory — an ability to compare what just happened to what happened earlier. How do plants remember?[10]

Darwin, who was a brilliant low-tech biological researcher, showed that carnivorous plants such as the Venus flytrap use a form of memory when deciding whether to shut their jaws. Darwin grew carnivorous plants in his own greenhouse, and in 1875 he published a pioneering book on the subject. He understood that carnivorous plants eat small creatures such as beetles or flies or small frogs because they live in nutrient-poor soils and need extra nitrogen and phosphorous. But they have to be discriminating predators because it costs energy to spring their jawlike traps and then reset them. So they have to be able to decide if something is worth trapping. Darwin could not trick carnivorous plants with drops of water or organisms so small they could escape. We now know that they spring their traps only if something touches at least two of a small number of tiny sensors inside their jaws, and the touches must occur in quick succession. The first touch says it is "possible" that something big has landed, get ready, and that information is remembered. The second touch says it is "probable," close your jaws!

Modern research shows that chemiosmosis is at work. The first touch triggers flows of calcium ions through the membranes of sensor cells, setting up an electric potential that is not quite enough to release the trap. The Venus flytrap is hesitating at the bookies. It will remember that first touch as long as the electric potential it created lasts. But the plant's enthusiasm for placing a bet will wane as the electric potential leaks away, unless it is renewed by getting new information quickly in the form of a new electric pulse. And that's exactly what a second touch of the trap will do. The two charges added together can generate an electric pulse just big enough to set off the trap. The memory consists in the addition of two charges within a short period of time, so it's a bit like a punter who receives two promising tips while heading for the bookies. In computational terms, this is a sort of "if A and B then C" switch. If two touches follow in quick succession, release the trap! After snapping its jaws shut, the plant secretes digestive juices, turning its jaws into what Darwin called "a temporary stomach."[11]

Plants use different mechanisms to create and store short-term and long-term memories. We have just seen a Venus flytrap deploying its short-term memory. Plants also have long-term memories. They can remember over hours or even months and years. Most plants detect seasonal changes by noting differences in the length of day and night.[12] But this won't tell a plant what season it is when days and nights are almost the same length. It also needs to know whether the temperature trend is rising or falling. And to detect a trend you need at least two data points, one of which must be stored in memory. Some plants know if they have recently lived through a cold or a warm period, and that helps them discriminate between fall (when they should probably be shedding leaves) and spring (when they should be growing them).

Crucial research on long-term memory in plants comes from studies of Arabidopsis because certain species need a cold spell before they can flower. Once more, epigenetic mechanisms are in play. In eukaryotic cells, DNA molecules are locked in the cell's nucleus, where they are twisted and packed tightly and economically around proteins known as histones that act like spools holding skeins of wool. Each bundle is packed within larger bundles to form tight balls of *chromatin*. When a gene needs to be expressed, transcription factors have to burrow through layers of chromatin, find the right strip of DNA, and unwind it before the gene can be read and expressed. So the way DNA is bundled up can determine how easy it is to get at and express particular genes. In plants such as Arabidopsis, periods of cold weather seem to ensure that the chromatin is folded to give easy access to the genes needed for germination.[13] But after germination, the histones will be repacked so as to block the expression of those genes until later in the year. This form of long-term memory arises from changes in the organization of the chromatin in which a cell's DNA is stored.

All organisms, including plants, seem to possess internal circadian clocks, which help them predict likely changes in the external world.[14] The rhythms of day and night are particularly important because they shape so many other rhythms on our planet, from temperature changes to the behavior of predators. It takes the eyes of some reef fish about twenty minutes to adjust to daylight. So about twenty minutes before dawn, their internal clocks say something like: "I know it still looks like night but there is a high probability of daylight in twenty minutes and an equally high probability that bad fish will start hunting you once it's light, so you'd better start waking up NOW!"[15]

The earliest evidence for the existence of circadian rhythms in plants was reported in the early eighteenth century by the French astronomer Jean-Jacques d'Ortous de Mairan. He noticed that the leaves of mimosa plants rose and fell depending on the position of the sun, and that they kept doing so even if placed in a dark cupboard, though over time their rhythms diverged from those of the sun. Clearly, their rhythmic behavior depended on some sort of internal clock. Some cyanobacteria can generate rhythmic cycles using just three proteins. In more complex organisms, there can be many different circadian clocks, which update and coordinate themselves in a process known as *entrainment.* Mammals have a sort of master clock in a part of the brain with the wonderful name of the *suprachiasmatic nucleus,* or

SCN. This coordinates many internal clocks, a bit like Greenwich Mean Time.[16] But no circadian clock is perfect. Plants suffer jet lag if their clocks are reset by a laboratory scientist cruel enough to artificially change the rhythms of day and night. Artificial sunlight can trick them into opening their leaves in the middle of the night. But unless they are dealing with a peculiarly malicious scientist who keeps changing the clocks, they will soon adjust to new rhythms. Today, lettuces, peas, zinnias, and sunflowers are being grown in the International Space Station in special environments that create artificial days and nights for them, and they adjust their circadian clocks accordingly.

The information-gathering and analytical skills of plants are so impressive that Darwin, in a rare flight of fancy, wondered if plants might not have some sort of brain, perhaps in the "radicle," or tip of a plant's shoots: "It is hardly an exaggeration," he wrote, "to say that the tip of the radicle . . . having the power of directing the movements of the adjoining parts, acts like the brain of one of the lower animals."[17] Today, this looks like one of Darwin's rare overstatements. The ability of plants to anticipate and plan for likely futures seems not to require a central coordinating system. Instead, it is distributed throughout the organism. The computational abilities of plants, like those of bacteria, seem to be an emergent property arising from the interactions of billions of individual biochemical reactions. But that should not reduce our respect for the sensory and computational skills that plants use to deal with uncertain futures, because so many of the future-management decisions taken by our own bodies arise in similar ways.

Finally, plants act. Sunflowers, mimosas, and many other plants turn their face to the sun; roots sniff out water and nutrients in the soil and burrow deeper to find them; different parts of a plant's body grow, shrink, change color, bud, or exude exotic scents. Many of these actions change the plant's body, but a surprising number involve movement. Plant motion so fascinated Darwin that he wrote a book called *The Power of Movement in Plants.* It showed that plants engage in elaborate balletic movements as they try to manage their futures. Many of these movements take the form of what he called "circumnutation." This is an exploratory circling motion, like "the stem of a climbing plant, which bends successively to all points of the compass, so that the tip revolves." Darwin found that "every growing part of every plant is continually circumnutating, though often on a small scale. Even the stems of seedlings before they have broken through the ground, as well as their buried radicles, circumnutate, as far as the pressure of the surrounding earth permits."[18]

Circumnutation shows how all the steps in future management are interlinked. It allows each part of a plant, whether it is a root or a twig or a leaf, to patrol its surroundings, dipping randomly into the flow of events, looking for opportunities or trends and clues as to what is about to happen. And circumnutation leads to action. The *Cuscuta,* or dodder plant, is a sort of vine related to morning glory. Like a vampire, it sucks the vital juices of its neighbors. Young dodder seedlings spiral randomly as they grow upward, sniffing the air for potential prey. They will identify possible victims if their sensor proteins lock on to specific chemicals in the air, just as our noses lock on to molecules drifting out of a decanter of good wine. If the dodder seedling sniffs "tomato," it will bend toward the tomato, wind around its stem, and bore into it with little drills. When the drills reach the phloem, which carries the tomato plant's sap, they will start sucking nutrients from their victim. The dodder will flourish while the tomato withers.[19]

However sinister the dodder's behavior may seem to us, the beautiful, twirling motions of circumnutation are a nice metaphor for the way that plants, and indeed all living organisms, cautiously spiral their way into the dark of their futures, taking random chances but using rules from their genomes, remembered clues from the past, and updates from their roots and leaves to place bets on likely futures.

How Animals Manage Their Futures Using Nervous Systems and Brains

Animals face more complex life challenges than plants because they move a lot. And they move because they get their nourishment by eating other organisms. Plants can wait for sunlight, rain, and nutrients to arrive on their doorstep. Fungi get off lightly too. They also eat other organisms, but unlike animals most fungi have the decency to wait until their prey are dead (though some fill their victims with psychotropic drugs, turning them into zombies, before consuming them alive).[20] Eating the dead makes life easier because the dead cannot run away, counterattack, or trick you. And as the remains of the dead are scattered generously around the Earth's surface, fungi, like plants, can usually find nourishment without moving much or thinking too hard. So, like plants, most fungi are sessile and can survive without the specialized computational systems that animals need.

The problem for animals is that most large organisms, including other animals, hate being eaten. (Grass and the fruit of some plants are rare exceptions, which may be why herbivores need smaller brains than carnivores.) Unlike the genetically engineered beef in Douglas Adams's *Restaurant at the End of the Universe,* which recommends its delicate flesh to diners before going off to shoot itself, most animals will run away or hide if you try to eat them, while many plants will try to stab you or sting you or poison you. So most animals only get to eat if they can outwit, outrun, or outmuscle their prey. They have to slither, crawl, climb, swim, or fly through the world in order to eat, and often they have to go on long and complex journeys. And when, finally, they've found a potential meal, they may have to fight it, or do some hard thinking to overcome its defenses.

In short, it is tough being an animal, and an animal's future is generally more diverse and unpredictable than that of a daisy or a mushroom. Like plants and fungi, animals need clear goals, lots of information, and a large repertoire of possible responses to rapidly changing situations. But it is the middle step of our three-pronged future-thinking process — assessing and analyzing trends in their environments — that poses the greatest challenge for animals. So the rest of this chapter will focus on the elaborate neurological machinery — the nervous systems — that allow animals to model likely futures with exceptional sophistication. What imagined futures flit through a young antelope's mind before it decides if it is safe to drink from the water hole? And how are those images created?

The Evolution of Nervous Systems

Animal nervous systems are built from networks of neurons: cells that specialize in efficient communication often over large distances. Neurons come in three main types. *Sensor* neurons detect information and *motor* neurons tell muscles what to do. Between sensor and motor neurons you find *interneurons*. Networks of interneurons analyze information from sensor neurons, compute likely futures, decide what to do, and pass their decisions on to motor neurons. In simple situations, or where there is no time to think, sensor neurons bypass the interneurons and send orders directly to motor neurons. You'll get the idea if you touch a red-hot iron and watch what your body does. It won't indulge in a lot of thinking. But in more complex situations, decisions are taken after they have been analyzed by networks of interneurons.

One measure of the growing importance to animals of careful future thinking is the increasing importance of interneurons in larger and more complex animals. The worm species *Caenorhabditis elegans* has just 302 neurons, and these are divided more or less equally among sensory neurons, motor neurons, and interneurons.[21] But as more elaborate nervous systems evolved, the proportion of interneurons increased, and more and more of them were concentrated in the special computational organs we call brains. The main task of brains is to think about and model likely futures with just the right balance between precision and generality. As the philosopher Patricia Churchland puts it, "Prediction . . . is the ultimate and most pervasive of brain functions."[22]

The earliest evidence for simple nervous systems appears from about six hundred million years ago, during the Ediacaran era, which is when the first animals flourished. Neurons have not changed much since then, but the computational power of nervous systems has increased by orders of magnitude as more and more neurons have been integrated within increasingly elaborate networks.

The simplest animals, such as sponges, don't have neurons or nervous systems. They don't need them because, like plants, they are sessile for much of their lives. Coelenterates, such as jellyfish, do possess neurons, but these are usually organized in networks with no central hub.[23] In some, however, such as hydra, neurons gather in rings near body regions such as the mouth or tentacles, where a lot happens.

More complex nervous systems evolved in bilateral animals, which have a front and back, a top and bottom, and a left and right side. And brains. Today, bilaterians account for most animal species, including worms and fish, lobsters and insects, crocodiles and humans.[24] Even in flatworms, we find neurons gathering in bulbs, or *ganglia*, toward the animal's front region, which is usually the first part of the body to encounter new trends. Many invertebrate species have several ganglia, which manage different parts of the body. Octopi, which are probably the brainiest invertebrates, keep most of their neurons in their tentacles. Arthropods, a vast group of animals that includes insects and crustaceans, have multipart brains formed from the fusion of two and sometimes three frontal ganglia.[25] Much of their work goes into managing the eyes, antennae, and mouth.

Nervous systems and brains have flourished most extravagantly in the lineage of vertebrates, or animals with spinal cords. The simplest way of appreciating the change is by counting the number of neurons in species alive today. As we have seen, the nervous system of *Caenorhabditis elegans* has only 302 neurons, so few that researchers have mapped all the connections between them. The sea slug, *Aplysia,* has about twenty thousand neurons. The fly species *Drosophila* has about two hundred thousand neurons in its brain, while honeybees, which are among the brainiest of insects, have about one million. Octopi can have as many as 550 million neurons.[26] Mammals have particularly large brains, and human brains may contain one hundred billion neurons, between which there may be up to one thousand trillion connections. Each neuron can send up to fifty signals per second, which means that the human brain can execute something like 10^{15} logical operations per second.[27]

Big brains are very good at detailed modeling of present realities and possible futures. The squishy ball of neurons between the ears of a thirsty young antelope can turn the millions of signals generated as it walks toward a waterhole into a moving, three-dimensional virtual image, complete with swaying, sweet-smelling grasses, buzzing insects, lots of other antelopes, and, yes, the scent and sight of a pride of lions patrolling the water hole. Damn! Not all these computations go on in the brain, of course. Many occur in networks of neurons that extend down the spine and throughout the body, which is why the antelope's legs are getting ready to run.

Vertebrate brains have three main parts: a forebrain, a midbrain, and a hindbrain, which links directly to the spinal cord. The midbrain and hindbrain manage processes over which we have no conscious control, such as walking and normal breathing. That means they take care of most future thinking. The forebrain can process more complex information and is particularly good at modeling possible futures, so it can play an executive role, adjudicating between recommendations from other parts of the nervous system.[28] The forebrain has grown particularly fast in the primate line of mammals. In the evolution of our own species, the part of the forebrain known as the neocortex grew spectacularly over just two million years. It almost tripled in size, and the part that grew fastest was the frontal cortex, which is thought to be the most important area for "working memory, action planning, and intelligence." Neurobiologist Gerhard Roth estimates that humans have about fifteen billion cortical neurons, while whales and elephants, our closest rivals in this respect, have only about eleven billion, and our closest relatives, the chimps, have about six billion.[29]

How Do Nervous Systems Work?

Nervous systems link neurons in huge networks, just as computers link electronic transistors. And like computers, nervous systems communicate (mainly) using electric pulses. Like transistors, neurons can receive and assess multiple electrical signals before deciding whether to pass them on. When linked together in large, well-designed networks, they can perform enormously complex computations and build rich models of the world. They can also store memories in networks of neurons that are held together for hours, days, or years.

To see how neurons go about their work, we will have to shrink ourselves back to the size of proteins once more and reenter the goo and slush of a cell's cytoplasm. Once again, we will be jostled and tugged by force fields and will keep bumping into proteins and other molecular citizens that are rushing about carrying out different tasks. But here things are on a larger scale than they were in the *E. coli* cell. We're in a eukaryotic city rather than a prokaryotic village; the population is more diverse and roams over greater distances.

It wasn't until the twentieth century that the basic structure of neurons was mapped by the Spanish neural biologist Santiago Cajal (1852–1934), who also made beautiful scientific drawings of the neurons he studied in his microscopes.[30]

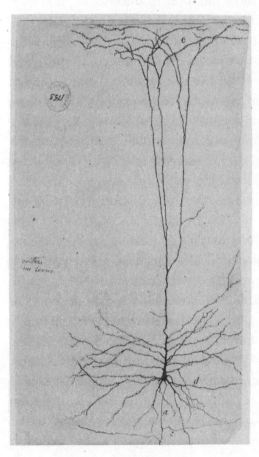

Figure 4.1: Drawing of a Giant Pyramidal Neuron of the Human Cerebral Cortex 1899 (ink and pencil on paper) The dark mass in the center is the cell body. The long dendrites in the upper part reach more than a millimeter to the surface of the brain (e). Other dendrites (d) are around the cell body. Look closely at the dendrites and you can see many synapses, which look almost like fur. The axon (a) of this neuron splits into branches (c).

Cajal showed that all neurons share the same three components. First, each has a main body that contains the cell's nucleus and much of its basic operating machinery, as well as organelles such as mitochondria, which supply energy. But it is the second and third components that make neurons different: *dendrites* and *axons*. These are two threadlike structures that extend from the neuron's main body to contact other cells. Information enters neurons through *synapses* on its many dendrites, travels to the cell body, and leaves through a single axon. Axons sometimes split into a small number of branches, and they can be very long by cellular standards. Human axons from the sciatic nerve reach from the base of the spine to the big toe.

In the 1920s, the biologist Edgar Adrian showed that neurons swap messages using electric pulses known as *action potentials*. These last just a few thousandths of a second and travel through axons to specific addresses, sometimes a long way away. All action potentials have about the same power and duration, varying only in their number and the speed of their repetitions. As Adrian put it, "All impulses are very much alike, whether the message is destined to arouse the sensation of light, of touch, or of pain; if they are crowded together, the sensation is intense; if they are separated by any interval, the sensation is correspondingly feeble."[31]

Generating action potentials uses a lot of energy that is provided by the ancient biochemical trick of chemiosmosis, first identified by the biochemist Peter Mitchell in the early 1960s. We have already met it. It's a process that has existed since life first appeared on Earth, and it is still at work in every cell of our bodies. All cells can maintain a slight difference in voltage across their membranes by pumping out positively charged ions, such as calcium or potassium, to create a negative internal charge.[32] That turns them into tiny batteries. Suddenly letting those positive ions flow back in creates the electrical spike of the action potential. But the constant pumping needed to maintain the voltage difference across a cell's membrane is hard work; it accounts for 80 percent of the energy used by the human brain. Every thought that flickers through your mind, every brilliant idea or painful memory, every model of your next date or job interview, is powered by chemiosmosis and made possible by the pumping of charged molecules across the membranes of millions of neurons.[33]

Hike out to an axon's tip, and you will reach a synapse where the axon meets — and passes information to — a dendrite on another neuron. There are two means by which information can be transferred across a synapse. It can be transferred quickly via an electric pulse, which works well when speed is more important than deliberation, as in your reaction to touching a red-hot poker. But signals can also be exchanged more slowly, and with more deliberation, through the movement of individual molecules known as neurotransmitters, which diffuse across the tiny gap at a synapse like hostage exchanges in a spy drama. Once across the synaptic gap, neurotransmitters generate a small pulse.[34] If the new pulse is negative, it will increase the negative charge inside the receiving neuron, reducing the chance of that neuron firing. If the charge is positive, it will increase the chance of the neuron firing. But the neuron will only fire — that is, pass a signal to another neuron — after summing up the *inhibitory* and *excitatory* pulses from tens or hundreds of other neurons, and only if they add up to a certain threshold.[35] This is a bit like a Venus flytrap deciding whether to close its jaws. Like the Venus flytrap, the neuron is weighing information from different sources before deciding whether to fire or not to fire.

Action potentials can carry information at about ninety feet per second. This is much slower than a modern computer. But the signals do not deteriorate, because they are sent in relays that maintain their strength over large distances, like phone messages carried through underground cables. That's why, when we stub our toes, the pain is not diminished on the journey from toe to brain.[36] Nervous systems also operate in parallel. At any one moment, huge numbers of action potentials are firing, collectively performing millions of simultaneous computations. Parallel computing explains why, in some ways, brains are still more powerful than the best computers.

How Do Nervous Systems Help Animals Think about the Future?

How does the firing of all these action potentials help animals think creatively and productively about likely futures? Arranged in vast networks, like transistors in a computer, neurons can collect information from our senses, analyze it, store it in memory, compare it to other memories, and interpolate missing information to build models of how the world is evolving. Think of catching a ball. You can remember when it was thrown and how fast; your mind will model its likely trajectory after interpolating information about its weight and momentum and the impact of the wind, and then calculate where your hand must be to catch it.

Memories of past trends are crucial building blocks for all models of possible futures. Nervous systems hold memories in more or less stable networks of linked neurons. Biologist Eric Kandel worked out how memories are formed by studying how neurons get locked together in the sea slug, *Aplysia*. Since his research, new imaging techniques such as positron-emission tomography (PET) and magnetic resonance imaging (MRI, or functional MRI, fMRI) have let researchers watch connections between neurons being made and unmade in real time by showing which parts of the brain light up when its owner is doing different kinds of thinking.

Learning and memories are held in networks of neurons

bound together as synaptic connections get reinforced. Short-term memories form quickly and cheaply and can be dissolved fast, like one-night stands, but neurons that meet many times can build more permanent relationships. Long-term memories, like marriages, require more investment but usually last longer.

We have already seen this distinction in plants. If a memory is required for just a few seconds, you don't want to waste energy on it. So short-term memories depend on cheap and reversible processes similar to the shape changing of proteins that we saw in *E. coli,* or the fading of an electric charge that we saw when the Venus flytrap was deciding whether to close its jaws. Laying down long-term memories requires more permanent changes to neurons, such as the addition or consolidation of new synapses, while forgetting may mean letting synapses atrophy.[37] To change a cell's anatomy in this way you need new working proteins, and that means getting transcription factors to burrow into the cell's DNA to activate genes that produce new proteins that build the new and more robust synaptic connections of long-term memories.[38]

Learning — something animals are exceptionally good at — means building new long-term memories or adjusting existing memories in response to new information about possible futures. Eric Kandel showed how the forging and breaking of links between networks of neurons can explain three basic forms of learning. Each can be thought of as an inductive prediction, because it makes a bet on likely futures given a particular trend in the past that has been stored in memory. *Habituation* is a sort of negative learning. It helps us unlearn something. It tells us that signal A will not always correlate with event B in the future. If you move to a house near an airport, you may be unsettled by an algorithm that says: "Sudden loud noises! Eek! Danger!" But you will soon learn that the roar of jet engines does not mean you are about to be attacked. *Sensitization* is the opposite.

It tells you that the correlation between signal A (touching a red-hot poker) and event B (lots of pain and hurt) is real. There is a real trend there, and it will probably continue into the future. Finally, in classical, or Pavlovian, conditioning, an organism learns to connect an arbitrary signal to a likely future outcome after repetition teaches it a new correlation and a new trend. The Russian physiologist Ivan Pavlov (1849–1936) rang bells before feeding dogs, and eventually the dogs began to salivate in expectation of a meal whenever the bells rang. (As a student I lived for a year in Leningrad / St. Petersburg, and while walking past the biology department of the University of Leningrad I often heard dogs barking. I was told they lived in labs once used by Pavlov, where his experiments were being repeated.) These three ways of learning about what is likely to happen seem to be present in some form in all living organisms, even single-celled bacteria.[39]

While habituation erodes synaptic connections, sensitization and Pavlovian learning multiply and strengthen them. This explains why the area of the frontal cortex devoted to the left hands of violinists (which control fingering) can be five times larger than the equivalent area in nonviolinists. Similar changes have been observed in the brains of London taxi drivers, in areas associated with spatial mapping.[40] Violinists and taxi drivers grow new synapses in specific areas of their brains once their nervous systems decide those patterns will help deal with the future. Their brains will invest in the necessary biochemical changes and new information will be learned and locked in long-term memory.

As they are updated, memories are used to build elaborate, plausible, and changeable models of the world. But often, as in a jigsaw with missing pieces, there are gaps, and that is where interpolation comes into play. We have seen from the example of the Necker cube how keen the mind is to build models from limited information. Our visual system illustrates how brains assemble new information into elaborate models of the world, helped by memories and a fair bit of interpolation and guess-work. Each eye has about one hundred million photoreceptor cells. Information from these cells is passed on to the brain, which processes it into different types of perception: colors, shapes, lines, angles. The mind then aggregates this informa-tion, correcting it with the help of memory, tidies it up, and interpolates missing information from memories of similar scenes, for example, to fill in the blind spot in the center of our eyes. The result is a rich and vivid model of what has probably been happening, a model whose trends can be used to antici-pate likely futures.

Some of these models are stored in our long-term memory and can be summoned up again and again. But memories, unlike photographs or written documents, are not simple dupli-cates. They are re-created each time they are recalled, so they change and are colored by later events. That is why, in retro-spect, remembered events can so easily be reinterpreted as pre-dictions. Two thousand years ago, Plutarch recorded many claimed predictions of Caesar's assassination. A seer had warned Caesar of "great peril" on the ides of March, and when, on his way to the Senate, Caesar "greeted the seer with a jest and said: 'Well, the Ides of March are come,' . . . the seer said to him softly: 'Ay, they are come, but they are not gone.'" Strabo reported other strange portents, including the ominous facts that Caesar himself had sacrificed an animal that turned out to have no heart and that his wife, Calpurnia, had dreamed of holding Caesar's dead body in her arms.[41] It is not surprising that we often remember predicting the future because we are constantly modeling possible futures, so the odds are good that some of those model futures will look like the futures that actu-ally turn up. They will look like predictions, in a phenomenon known to forecasters as "hindsight bias."[42] Particularly with a bit of retrospective massaging of our memories! No wonder retrospective predictions have been "remembered" for many modern events including the 9/11 attacks on the United States and the global financial crisis in 2008.

To us, the memories and models put together by our minds *are* the world. They have the color, drama, and zing of reality. Indeed, they are as close as we can get to reality. As social psychologist Daniel Gilbert puts it, our brain

> gathers information, makes shrewd judgments and even shrewder guesses, and offers us its best interpretations of the way things are. Because those interpretations are usually so good, because they usually bear such a striking resemblance to the world as it is actually constituted, *we do not realize that we are seeing an interpretation.* Instead, we feel as though we are sitting comfortably inside our heads, looking out through the clear glass windshield of our eyes, watching the world as it truly is. We tend to forget that our brains are talented forgers, weaving a tapestry of memory and perception whose detail is so compelling that its inauthenticity is rarely detected.[43]

The models our minds create are really, as the consciousness researcher Anil Seth says, "controlled hallucinations."[44] They are the mind's best predictions of what is probably out there, given the signals it is receiving. They are as close as we can get to what is really happening because, directly or indirectly, they are based on a lot of information that we have gathered about what is out there. These models are our windows onto the world and the future, and they shape every aspect of our future thinking.

Of course, our future-anticipation machinery consists of more than fleshy neuronal computers. In brainy organisms such as mammals, algorithms or rules of thumb that have worked well in the past are reinforced by emotions. Our young antelope friend's brain and body don't just tell him that it would be smart to run away from the lion pride. They also secrete hormones that energize him like fire alarms by creating powerful emotions of fear and panic to get him running. They take him into the Red Zone of Future Cone 3. It's the same with humans. Our emotional systems are part of a large repertoire of semi-automatic responses to familiar situations, situations in which careful thought is probably not necessary but rapid reactions are. Just think of the panic you feel when you cannot breathe. The link between nervous systems and emotions explains why the distinction between good and bad futures is not just thought about; it is also felt — strongly. We have powerful feelings about many of the things that we and others do, and for humans at least, those feelings lay the foundations for much of our moral and ethical thinking.

Emotions are closely linked to quick algorithms that give us advice about the future based on familiar trends rather than on rigorous thinking about what is actually going on. These handy algorithms are part of what the psychologist Daniel Kahneman calls "fast thinking."[45] Fast thinking is intuitive, it mostly takes place below consciousness, and it requires little deliberate effort. But it handles most of our decisions about the future. It is indispensable when there is not enough time or information or energy to think a problem through deeply; it is cheap and easy in simple situations; and usually fast thinking points us in the right direction. But not always. Sometimes, fast thinking is too fast, as the research of Kahneman and his partner, Amos Tversky, showed. For example, fast thinking relies on information that is readily available, and that often means it draws conclusions prematurely, based on what Kahneman and Tversky call, facetiously, the "law of small numbers."[46] Our young antelope might say to its mother: "I went to the water hole four times, and there were lions there, but they didn't try to kill me. Why don't you want me to go there?" We all generalize from recent experience, and often that means from ridiculously small samples, like the sports franchise that sacks its coach after two unsuccessful seasons. Fast thinking explains why so much of our future thinking is dangerously ad hoc, even when powered by sophisticated nervous systems.

On the other hand, if fast thinking gets us into trouble, and we have enough time and mental energy, species with large brains can mobilize a second system that Kahneman calls "slow thinking." This is the thinking used by the mother antelope who replies, "Your sample is far too small to generalize from. I've been around longer and remember your dad's last day. STAY AWAY FROM THE WATER HOLE!" Slow thinking requires some level of consciousness, so this is when we can begin to talk more literally about *future thinking*. Slow thinking requires more effort and focus than fast thinking, but it works through problems more rigorously, uses more information, and checks its conclusions more carefully. So, in large, brainy organisms like us, it is slow, conscious thinking that makes many of the big executive decisions about the future. The division of labor between fast and slow thinking works well most of the time. As Kahneman puts it, "It minimizes effort and optimizes performance."[47]

Finally, much remains mysterious about the biology of future thinking. As the computer scientist Stuart Russell puts it (with only slight exaggeration), how neurons do "learning, knowing, remembering, reasoning, planning, deciding, and so on — is still mostly anyone's guess."[48] We have no real idea how trillions of neurons holding hands can generate the vivid *feeling* of reality that we carry in our heads as we prepare for the future. We have no real understanding of the source of awareness or consciousness (which Alison Gopnik defines as "that thing that anesthetics get rid of").[49] Nor do we know at what point in evolution the glow of consciousness first arises. Is the thirsty young antelope thinking consciously about his future fate? Many philosophers describe the problem of consciousness as "the hard problem," a label first used by the philosopher David Chalmers in 1995. Consciousness is as problematic for many philosophers and psychologists as dark matter and dark energy are for cosmologists. But whatever consciousness is, it allows the awareness of future thinking that is present for much of our waking lives.

PART III

Preparing for Futures

How Humans Do It

CHAPTER 5

What Is New about Human Future Thinking?

But Mousie, thou art no thy-lane [on your own]
In proving foresight may be vain:
The best laid schemes o' mice an' men
Gang aft agley, [often go astray]
An' lea'e us nought but grief an' pain,
For promis'd joy!

Still, thou art blest compar'd wi' me:
The present only toucheth thee,
But och! I backward cast my e'e [eye]
On prospects drear —
An' forward, though I canna see,
I guess an' fear!
— ROBERT BURNS, "TO A MOUSE, ON TURNING
HER UP IN HER NEST WITH THE PLOUGH"[1]

In humans, and probably in many other brainy species, including Burns's "Wee, sleeket [glossy], cowran', tim'rous beastie," a lot of executive-level future thinking is conscious. But two linked changes have given human future thinking unprecedented power and significance, even in comparison with other brainy species. As our species evolved over several million years, neurological and biological changes allowed individual humans to think about, imagine, plan, and model possible futures with exceptional virtuosity. But the impact of these skills was magnified many times over by a second change: the development of human language, which let humans share ideas and accumulate information collectively. Sharing information among many individuals meant that human future thinking and management, along with human technologies and human culture in general, could evolve and change faster than ever before, becoming more powerful from generation to generation over several hundred thousand years. Together, these changes have revolutionized the relationship of our species to the future and to the planet that is our home. We have become, in the lovely metaphor of Didier Sornette, "Dragon-kings," known creatures that have suddenly started behaving in fabulous new ways.[2]

The Biological Differences

We belong to a group of exceptionally brainy bipedal primates called *hominins*, which evolved in the past few million years. In the last two million years, hominin brains expanded at warp speed. The size of modern chimp brains ranges from 300 to almost 480 cubic centimeters. Two million years ago, the hominins known as *Homo erectus/ergaster* had brains of about 900–1,000 cubic centimeters.[3] Modern human brains are about 1,300 to 1,400 cubic centimeters, and those of our Neanderthal cousins were even larger, ranging up to more than 1,500 cubic centimeters.

Brain size is not everything, of course. The largest known brain, that of a sperm whale, is about 8,000 cubic centimeters. More significant is the ratio between brain size and body size, because large organisms have more neurological real estate to

manage. That is why, in evolutionary history, brains have tended
to expand as bodies got larger. But hominin brains expanded
faster than this rule predicts. Human brains are exceptionally
large when compared to human bodies.[4] As we saw in the previous chapter, humans also have an exceptional number of neurons in their frontal cortex, the region that specializes in
computation and planning.

What drove these changes? Evolutionary biologists have to
ask because the constant firing of action potentials by billions
of neurons guzzles vast amounts of energy that could be used
for other jobs. That makes large brains expensive, which is
why, in evolutionary terms, they are rare. There must have
been powerful reasons for the evolution of large brains. The
speed of the change (on evolutionary time scales) suggests that
positive feedback loops were at work. One possible loop is
between brain size and sociability. Mammals are warm-
blooded. To maintain their body temperatures they need up
to ten times more food per gram of body weight than reptiles.
One way to get it is to be more cunning; another is to cooperate.[5] So perhaps it is no surprise that mammals tend to have
large brains and many live in herds or packs that can pool
skills and muscle power. But living in groups is intellectually
demanding, because you can't think just about your own
future.[6] You also have to think about the futures of others. You
acquire debts and obligations and have to keep track of them.
You have to guess what the alpha female is thinking, or what
your enemy may be plotting.[7] So sociability probably encourages larger brains, while larger brains enable sociability — this
is a powerful feedback loop.

However we explain the rapid expansion of hominin brains,
it revolutionized human future thinking. The frontal cortex is
usually thought of as the seat of working memory, and the areas
that expanded fastest manage our sense of temporal change,
our emotions, and our sense of purpose and planning. The

same areas also help integrate visual and other sensory information to create models of our environment and arrange imagined events along imagined timelines: just the skills you need to model alternative futures. As Patricia Churchland writes, a larger prefrontal cortex means "greater capacity to predict, both in the social and in the physical domain."[8]

And humans really are better at modeling sequences of likely future events, such as those required to make a complex stone tool or to manage fire. Humans are exceptionally good at imagining futures on large time scales — a feature we will explore in spectacular forms in the final chapters of this book. More room for thought also means more room for *careful* thought, the ability to think things through with deliberation and focus, to shift from "fast" to "slow" thinking.[9] When not rushed, brains can choose to focus on parts of the flood of information streaming through them, and humans seem to be exceptionally good at focusing attention, even in the presence of distractions. This is the skill that meditators practice. Focused, conscious thinking increases our ability to think through complex chains of reasoning or compare possible futures.

In short, human brains seem to have paid for their passage by enhancing the ability of our species to imagine, mull over, and compare many possible futures.

The Social and Cultural Differences: Language and Collective Learning

Along with those enhanced skills our lineage got an unexpected evolutionary bonus. Larger brains enabled a second, even more transformative change: collective learning (or cultural evolution). Many species have some form of culture because they have languages and can share information and ideas. But humans are unique in sharing information so precisely and on such a scale that collective stores of knowledge grow and evolve across generations until they have transformed our place in the world. This is what I mean by "collective learning."[10] Collective learning explains why our species has a history because, as collective knowledge has accumulated, our technologies, lifeways, and ways of thinking have changed profoundly, very slowly at first and then more rapidly, as larger stores of knowledge gave us increasing power over the landscapes and organisms all around us.

Collective learning accelerated change. The great drivers of planetary history — plate tectonics, the rotation of the sun and moon, and evolution by natural selection — work mostly on scales of thousands or millions of years. But collective learning works almost instantaneously, as people whisper ideas and pass them on. Natural selection still plays a role in human history, of course. It explains why groups descended from pastoralists can usually digest milk even as adults. But it is collective learning that has made us so different from all other species on Earth, on time scales much faster than those of natural selection. Alex Mesoudi, a specialist in cultural evolution, describes the accumulation of knowledge over many generations as "the defining characteristic of human culture."[11]

Collective learning and cultural evolution were made possible by the evolution of human language, which connects humans within what linguist Steven Pinker calls "an information-sharing network with formidable collective powers."[12] We don't fully understand how human language evolved, though there are many promising hypotheses. The same synergies that linked brain growth and sociability may also have driven the evolution of human language, as increasing sociability encouraged better communication between individuals eager to know what others were thinking and planning.[13] No wonder all social species, including birds, whales, and primates, have some form of language. Baboons can warn each other of danger with simple messages, such as "Look out! Eagle incoming!" But human language is exceptionally powerful. With more space in their frontal cortex, humans have the neurological room for vast stores of names, words, and concepts, and also for the grammatical workbenches and lathes on which we turn words and concepts into stories about real and hypothetical worlds.[14]

Whatever its origins, human language led our species across a fundamental threshold both as individuals and as a species. The psychologist Lev Vygotsky argued that words offer each one of us new ways of modeling our surroundings because they compress so much information.[15] Think of the explosion of thought inside your head when someone says, "Pink elephant!" Four syllables send signals ricocheting through networks of neurons to light up a vivid, complex idea in your mind. What's more, the idea is of something you have never seen before, at least not in the real world. Grammar can arrange the idea packages of words and phrases into complex stories that we can turn and twist in new configurations to model alternative futures. And we can do all that in the safe hypothetical workrooms of our minds, rather than taking our chances in the real world. Learning to talk turbocharges the future thinking of infants by helping them model possible futures in play.[16]

But the most dramatic impact of language has been on our collective learning and thinking rather than our individual thought processes. Language allowed each of us to dip into and contribute to the vast and growing pool of knowledge accumulated from generation to generation in all human communities. Shared stores of well-tested knowledge gave humans exceptional power over their surroundings and over other species of animals and plants, which is why all human communities have valued and nurtured traditional knowledge. For most of human history, knowledge was stored in rich oral archives, in songs and stories, in monuments and landscapes. Traditional knowledge was taught and passed on carefully, often through ritual, and it could take many decades to master fully. Collective learning also tested ideas, as natural selection tests possible species, because unsuccessful or outdated ideas were likely, sooner or later, to be "falsified," in the jargon of the philosopher of science Karl Popper.[17] As early as the eighteenth century, Adam Ferguson, a friend of Adam Smith and David Hume, understood how transformative these changes were: "In other classes of animals, the individual advances from infancy to age or maturity; . . . but, in the human kind, the species has a progress as well as the individual; they build in every subsequent age on foundations formerly laid."[18]

Collective learning unleashed powerful trends that would take human history in new directions. For most of human history, these trends worked so slowly they could not be seen. What stood out were cyclical patterns — the rise and fall of individual families or communities or empires. But looking back from today, armed with much more knowledge of the past, it is easier to see that collective learning also unleashed trends that would shape all of human history. Three trends stand out. First, collective learning made humans more and more powerful. A never-ending trickle of new ideas and technologies gave humans increasing power to control their environments and manage their futures. Second, collective learning shared ideas on increasing scales as human networks grew larger, eventually spreading across continents until today we exchange ideas, goods, and people within a single global network of phenomenal power. Expanding human networks also made it possible for humans to situate themselves within expanding "circles of concern," as their sense of community embraced larger and larger groups, such as the tribe or the nation. Third, collective learning accelerated the pace of change because it generated so many feedback loops — innovations that encouraged more innovations. For most of human history, change was so slow that it was hard to see. Only in recent millennia, and particularly in recent centuries, has change become so rapid that it seems inescapable. "In the past," wrote the philosopher Alfred North Whitehead, "the time-span of important change was considerably longer than that of a single human life. Thus mankind was trained to adapt itself to fixed conditions. Today this time-span is considerably shorter than that of human life, and accordingly our training must prepare individuals to face a novelty of conditions."[19] That is why the modern sense of time has the turbulence of A-series time rather than the serenity of B-series time.

These three trends — increasing technological power, expanding networks of exchange, and accelerating change — help explain why humans' understanding of time and the future has changed so profoundly in the course of our history.

The Archaeology and Anthropology of Time: Why Experiences of Time Differ

It is extremely difficult to track how the future thinking of our species has changed over the course of human history because for most of the hundreds of thousands of years since humans first appeared, ideas left few traces. It would be wonderful to send a team of anthropologists back in time with recorders and cameras and universal translators (perhaps one of the Babel fish that you stick in your ear in Douglas Adams's *Hitchhiker's Guide to the Galaxy*), to interview our ancestors. But it won't happen. Written evidence would be wonderful too, perhaps just a journal or diary or philosophical tract from fifty thousand years ago. Sadly, the earliest written documents are no more than five thousand years old.

Lack of evidence forces us back to Nasreddin Hoca's

research strategy of searching where the light is. Here, this means studying modern hunter-gatherer societies or reading the anthropologists who have tried to describe their worlds, in the hope that those societies have preserved traces of ancient ideas about time and the future. But how likely is it really that the modern peoples of the Kalahari, or indigenous Australians, or modern Arctic peoples still hold ideas that have anything in common with the most ancient forms of future thinking? The truth is that we do not really know, and many anthropologists are skeptical.[20] Still, the diverse palette of attitudes to time revealed in modern anthropological research encourages tentative speculation about ancient attitudes to the future.

The Anthropology of Time

In the 1940s, the American linguist Benjamin Whorf argued that some societies have no sense of time. In the Hopi language he found "no words, grammatical forms, constructions, or expressions which refer directly to what we call 'time.'"[21] At about the same time, the Romanian philosopher of religion Mircea Eliade argued, using a combination of archaeology and modern evidence about traditional societies, that people thought very differently about time in the small-scale societies of the past.[22] They did not see time as dynamic and linear, like the clock time of today's world. Instead, Eliade claimed, they experienced time in two linked ways: "profane time" and "sacred time." Profane time was the surface experience of change, a realm a bit like A-series time except that most change was seen as repetitive and cyclical, as in the setting of the sun or the coming of winter or the cycles of life, birth, and death. Sacred time was a bit like B-series time, a time of permanence and stability that could be accessed through ritual, in dreams, in sacred knowledge, and in trance. In Eliade's view, glimpses of sacred time convinced people that change was an illusion. At large scales, little changed.

In the twentieth century, anthropologists realized that our modern sense of a single, dynamic, onward-driving clock time arose recently, and that may be why some scholars, including Whorf, abandoned the idea of time as a fundamental category of human thought. Today, though, most anthropologists accept that beneath diverse attitudes to time, there *are* important commonalities. Whorf's claims about the timelessness of Hopi languages are no longer accepted, because later research showed that even languages without grammatical tenses have ways of dealing with pasts and futures.[23] As the anthropologist E. E. Evans-Pritchard wrote, among the Tiv peoples of northern Nigeria, there seemed to be no distinct category equivalent to the modern idea of time, and yet "time is implicit in Tiv thought and speech."[24]

Similarly, many anthropologists today would argue that Mircea Eliade exaggerated the uniqueness of ancient experiences of profane and sacred time. As we saw in chapter 1, even modern ideas about time include both dynamism and a sense of permanence. Cyclical and repetitive patterns of change are familiar in modern lifeways, and the sense of a deep permanence lurking beneath surface changes is alive and well in the laws of modern physics and in modern philosophies of time.

In a 1992 summary of the anthropological literature on time, Alfred Gell writes:

There is no fairyland where people experience time in a way that is markedly unlike the way in which we do ourselves [today], where there is no past, present and future, where time stands still, or chases its own tail, or swings back and forth like a pendulum. . . . There are only other clocks, other schedules to keep abreast of, other frustrating delays, happy anticipations,

unexpected turns of events and long stretches of grinding monotony.[25]

The anthropologist Jack Goody argues that all humans
experience temporal sequence (some events follow other events)
and temporal duration (some events take a long time; others
happen fast).[26]

Why, then, have anthropologists found such diversity in the
way different communities have experienced and described
time, the past, and the future?

Natural, Psychological, and Social Time

One way of explaining this diversity is to see that human experiences of time blend three distinct types of rhythms: the rhythms
of *natural time, psychological time,* and *social time.* To survive, we
have to align our own activities with these rhythms. But as
human societies and technologies have changed, the relative
significance of these rhythms has changed, transforming how
different societies experienced and understood time and the
future.

Natural time beats to the rhythms of day and night, rainy
and dry seasons, the motions of the sun, stars, and planets.
These rhythms, in particular the circadian rhythms of day and
night, shape the lives of all living organisms. In everyday life, it
is the repetitive aspects of natural time that stand out, the swing
from day to night, summer to winter, high tide to low tide. Only
in recent centuries have we learned how to track the longer and
more linear rhythms of natural time over thousands or millions
of years, such as long-term climate change or the tectonic
motions of continents and oceans. For most of human history,
natural time seemed to consist of endless, repetitive cycles, like
those of Eliade's profane time.

Psychological time is mercurial. It follows the rhythms of our bodies, the ebb and flow of hormones, breath, beating hearts, hunger and satiation, waking and sleeping, excitement and tedium, terror and contentment. These rhythms are dominated by immediate experience. They can pulse metronomically, like heartbeats. They can change in a flash when we panic. They can speed up and slow down. When we are bored, time crawls. (You can test this claim by staring at the second hand of a watch for five minutes.) As we age, time speeds up, so that birthdays and tax time seem to arrive faster each year. This may be because our life span is the yardstick against which we measure our inner experience of total time. For a year-old baby, one year is all of time; for a centenarian it is one-hundredth of time. We have some control over psychological time. Intoxication, prolonged dancing, trance, excitement, and stillness can alter how it works. Meditators sometimes report a stillness so deep that the sense of time's flow vanishes. Modern literature captures the protean rhythms of psychological time in the inner monologues penned by James Joyce, or Virginia Woolf, or Marcel Proust.[27]

Social time is built from rhythms created by other humans to which we must align our own behavior. It exists for all social creatures. But for humans, over the course of history, it has become increasingly powerful as once independent communities have found themselves enmeshed in the gears and cogwheels driving larger and larger networks of exchange. Today, social time regularly overrides our sense of natural or psychological time. Fly from Sydney to London and arrive at 8 a.m. (I've done it several times.) Your body will tell you it's time to go to bed, but if you're on a tight schedule, social time will tell you the day is just starting, and you'll probably force yourself to go with social time. In the enmeshed societies of today's world, we have to align our activities to those of millions of others.

The call to prayer, tax time, school bells, and calendars all help to instill a sense of social time. Our sense of social time precipitates from conversations and timetables, from rituals and schedules, and from our many social and legal obligations.

In a pioneering work on changing experiences of time, the sociologist Norbert Elias argues that the increasing scale of human networks of exchange has been the primary shaper of our modern sense of time because it has increased the power of social time. Large networks lock you into rhythmic grids shaped by millions of others. And as these grids get larger, they also shape our sense of the past and the future.

> Just as the chains of interdependency in the case of pre-state societies are comparatively short, so their members' experience of past and future as distinct from the present is less developed. In people's experience, the immediate present — that which is here and now — stands out more sharply than either past or future. . . . In later societies, on the other hand, past, present and future are more sharply distinguished. The need and the capacity to foresee, and thus considerations of a relatively distant future, gain stronger and stronger influence on all activities to be undertaken here and now.[28]

It was the sociologist Émile Durkheim who first persuaded many scholars that society shapes much of our experience of time.[29] Like Kant, he saw our sense of time not as a property of the universe, but as a sort of projection on the world. But for Durkheim the projection is collective rather than individual, social rather than psychological. It is instilled in us by the distinctive rhythms of the communities we grew up in and today it often overrides the rhythms of nature or our psyches.[30]

A Speculative Model of Future Thinking in the Foundational Era

With the help of these simple ideas about what has shaped our changing sense of time, we can try to build a speculative model of future thinking in the earliest human societies.

The earliest periods of human history, from the evolution of *Homo sapiens* several hundred thousand years ago to the end of the last ice age about ten thousand years ago, are often described as the "Stone Age" or the "Paleolithic era." But here I will describe them as the *foundational era,* because this is when the social, cultural, technological, and moral foundations were laid down for the rest of human history. Archaeological and anthropological research suggests that most societies of the foundational era consisted of small groups the size of extended families. People nomadized through intimately known home territories and hunted and gathered for subsistence using constantly evolving technologies handed down through the generations and finely tuned to their environments. Technological innovations and the pressure of changing climates encouraged small groups to enter new environments, from tropical forests to Arctic tundra, with the result that, over several hundred thousand years, humans slowly spread to all continents on Earth apart from Antarctica.

Ten thousand years ago, at the end of the foundational era, there were probably fewer than six million humans on Earth, but they could be found all the way from Southern Africa to Siberia and throughout the Americas.[31] In contrast to today's citified world of eight billion people divided into almost two hundred nations, the foundational world consisted of tens of thousands of tiny communities, each with its own home territories, traditions, and technologies, and each in contact with only a few neighboring communities. Like modern hunter-gatherers, ancient communities probably met with neighbors once or twice a year to exchange gifts, ideas, ritual practices, stories, knowledge, people, and genes. These ancient equivalents of the Olympic Games meant that each individual might meet several hundred people in a lifetime, and encounter a variety of cultural, ritual, technological, and linguistic traditions. But large gatherings were possible only in times of abundance, when many could feed from a small area, such as during the salmon runs of the North American Northwest, or the migrations of deer in southern France at the end of the last ice age.

Though information was exchanged between neighboring groups, it was local knowledge that mattered most, which is why the societies of the foundational era were extremely diverse. Accumulated and tested over many generations, passed on through repetition in stories, songs, and rituals, and occasionally traded to neighbors, local knowledge was practical, empirical, detailed, precise, and, in many senses of the word, scientific. In Australia before European colonization, writes the anthropologist Deborah Bird Rose, "the basic element to subsistence was neither technology nor labour, but knowledge." That knowledge included "resource locations, water sources, ecological processes, types of landforms, seasonable variability, animal behavior, cycles of growth, and types of plants and animals suitable as technical items, foods, medicines and 'tobacco.' Much of the knowledge was coded in song and story."[32]

Knowledge was probably the main source of social power in the foundational era. There were few differences in wealth or coercive power, but all small-scale societies had bodies of arcane knowledge accessible only to particular individuals, and that could magnify differences in authority and power.[33] It may be that special knowledge of the future was also restricted in this way.

Ideas about Time and the Future in the Foundational Era

Our speculative reconstruction of foundational-era thinking about time and the future will concentrate on four main features. (1) Communities were small and links were personal, so the future, too, was personal. (2) Our ancestors thought of the world they inhabited as a place they were part of, with its own laws that had to be respected. No modern hunter-gatherer societies show the hubristic modern sense of the future as a realm to be manipulated to satisfy human needs. (3) It seems likely that most people thought of the world as fundamentally stable despite surface changes. Change there certainly was, and it could be catastrophic. But most change was personal and cyclical, and beneath it lay a Parmenidean world of deep permanence, in which the future was expected to be much like the past. (4) Most people experienced the world as full of spirits and beings and forces that could shape both the present and the future. They could be negotiated with or fought, like all purposeful beings, and those relationships shaped many aspects of future thinking and planning.

First, in small communities, the futures that mattered were those of the people, the animals, and the plants of your homeland. The local weather mattered, the success of a seal hunt, the gathering of yams or tubers, relations with neighbors, health and sickness, the flourishing of prey species and edible plants, the life cycles of those around you. The future was personal, unlike the global futures we will discuss in chapter 8.

Second, the limitations of foundational-era technologies ensured that people had a sense of living *in* and *with* the world, rather than of dominating it or transforming it. All modern hunter-gatherer societies seem to have shared a powerful sense that there are universal ecological and moral laws that require humans to protect and care for the land. Humans could and did affect the mix of local plants and animals by burning the land or driving some species to extinction by overhunting. Indeed, most landscapes of the foundational era were shaped significantly by human activities. But the limits of human activity were well understood, and many stories told of the punishments for neglect or disrespect of "the law." It was foolish to neglect the rituals that kept things going. It was foolish to kill the young of a favored prey species or burn the land heedlessly. In the early 1950s, anthropologist Elizabeth Marshall visited and lived with the Ju/wasi, a people of the Kalahari who lived by hunting and gathering. Unlike most agrarian peoples, the Ju/wasi showed little interest in trying to manipulate or control their world: "People did not try to force the natural world. They did not, for instance, try to make rain, or make animals fertile, or make plants grow. With the exception of burning dry grass now and then in order to encourage green grass, they did not try to make things happen."[34] In the mythologies of hunters and gatherers, there is no equivalent to the widespread modern sense that humans are separate from and rule over the natural world.[35] Instead, people thought of themselves as *part* of the world, so they aligned their activities to the rhythms of natural and psychological time, rather than imposing their own rhythms on their surroundings.

Social time mattered, of course, and could sometimes override the rhythms of natural or psychological time. All communities probably used astronomy or changes in their environment to construct calendars for social activities and rituals. Indeed, the American archaeologist Alexander Marshack has argued that objects thirty thousand years old might be early forms of calendars.[36] But social rhythms did not dominate the sense of time as they do today. Here is anthropologist Richard Lee's description of the rhythms of daily life in the societies he studied in the Kalahari in the 1960s.

A woman gathers on one day enough food to feed her family for three days, and spends the rest of her time resting in camp, doing embroidery, visiting other camps, or entertaining visitors from other camps. For each day at home, kitchen routines, such as cooking, nut cracking, collecting firewood, and fetching water, occupy one to three hours of her time. . . . The hunters tend to work more frequently than the women, but their schedule is uneven. It is not unusual for a man to hunt avidly for a week and then do no hunting at all for two or three weeks. Since hunting is an unpredictable business and subject to magical control, hunters sometimes experience a run of bad luck and stop hunting for a month or longer. During these periods, visiting, entertaining, and especially dancing are the primary activities of men.[37]

In ancient hunter-gatherer communities, people seem to have lived comfortably with the many different rhythms of the surrounding world, the rhythms of dreams and the body, of gathering or hunting, of the sun, the moon, and the tides, of animal migrations and community rituals. This is very different from today's world, in which a uniform clock time creates a single grid for most of the rhythms that surround us.

Third, there are many hints that despite living, like all of us, in the flow of A-series time, people *thought* of time as fundamentally stable — more like B-series time. Beneath the changes of daily life and personal experience, people imagined a stable, mostly unchanging Parmenidean world. That could explain why many small-scale societies showed little interest in detailed historical timelines. Lynne Kelly, a scholar of oral cultures, writes of concepts of time among the Yolngu in Australia's Arnhem Land: "Time is not chronological. Mythological events are considered as occurring in the distant past and also as part of a continuous present."[38]

Anthropological evidence suggests that the past was not imagined as a single timeline moving further and further from the present. Instead, the past faded away quite rapidly into a misty era of beginnings, the "dreaming," to borrow the metaphor often used to describe indigenous Australian beliefs about the ancestral past. The word *dreaming* translates a term used by the Arrernte people, who live near Alice Springs. But the translation is misleading. As the anthropologist Roslyn Haynes puts it, the original term refers to "an ever-present reality, a dimension more real and fundamental than the physical world, which is merely temporal and contingent."[39] The Australian anthropologist W. E. H. Stanner described that realm as "everywhen." The Arrernte word translated as "dreaming" can also mean "the Law," the way things are and have to be and always will be. This is close to the idea of the *dharma,* a fundamental concept in many Indian traditions. The historian Ann McGrath writes, "In many Aboriginal languages, there is an expression to convey the concept of 'long, long ago' — a zone that also converges with the 'dreaming,' creation-time, which is actually not a discrete time at all, but an ongoing process."[40] The past is seen not so much as a continuum, but as a reservoir of knowledge and truth for the present.

In a Parmenidean universe, the past and future lose their distinctiveness and significance. It is the present that matters. In traditional Australian thought, what you really needed to know was *where* you were rather than *when* you were.[41] What is your country? Things and loyalties and stories and knowledge originated from places rather than from time. In such a world, maps are more important than timelines. Jack Goody writes:

> In nonliterate cultures ideas and attitudes concerning the past tend to reflect present concerns. To some extent this happens in all societies, especially in those situations where we rely on memory. But where the

transmission of culture is entirely dependent upon oral communication, . . . the past is inevitably swallowed up in the present. . . . Before (and partly after) the widespread use of writing, the past is a backward projection of the present, going straight back to the mythical age that saw the emergence of humanity and its present way of life.[42]

Something of this Parmenidean temporal sensibility survives within literate cultures, as in the beautiful verses from Ecclesiastes:

One generation passeth away, and another generation cometh: but the earth abideth for ever.
. . . .The thing that hath been, it is that which shall be; and that which is done is that which shall be done: and there is no new thing under the sun.
Is there any thing whereof it may be said, See, this is new? it hath been already of old time, which was before us.
There is no remembrance of former things; neither shall there be any remembrance of things that are to come with those that shall come after.[43]

In a Parmenidean universe, the future need not seem mysterious or threatening, because little changes beneath the surface ripples. This may explain something that puzzled anthropologists in the late twentieth century: modern hunter-gatherers seem to waste little time worrying about the future. A whole generation of anthropologists showed that what some observers interpreted as laziness or irresponsibility arose from an experience of the world as known and familiar. It would provide in the future as it had in the past.[44] "Why," the Khoisan people of the Kalahari Desert asked the anthropologist Richard Lee, "should we plant, when there are so many mongomongo nuts in the world?"[45]

Having said this, short-term futures on scales of weeks, months, or a few years were always important, and basic survival required an ability to predict the birth of a baby, the migrations of deer or kangaroo, or when to harvest mongomongo nuts. At this level, prediction was as pragmatic, empirical, and trend based as in all known societies. Astronomy was a reliable indicator of annual cycles everywhere, as well as a crucial aid to navigation, which is why all known societies have valued astronomical knowledge. As a European arrival in Australia noted in the nineteenth century, indigenous knowledge of the skies "greatly exceeds that of most white people. Of such importance is a knowledge of the stars to the Aborigines in their night journeys, and of their positions denoting the particular seasons of the year, that astronomy is considered one of the principal branches of education."[46] Astronomy was surely important for all foundational-era societies, but what it taught was how little changed beneath the surface.

Fourth, much of the future thinking of the foundational era was probably shaped by the assumption that the world is full of spirits and occult forces, just at the edge of perception. Most known religious traditions have assumed the existence of spirits and gods. Two thousand years ago, Cicero wrote, "Most thinkers have affirmed that the gods exist, and this is the most probable view, and the one to which we are all led by Nature's guidance."[47] These beliefs explain a distinctive type of future thinking, present in most human societies, and based on the conviction that you could inquire or negotiate about the future with beings from the spirit world, just as you could with other humans.

There may be neurological reasons for the ubiquity of such beliefs. All humans, and perhaps many other brainy species, distinguish between the living and the nonliving, between agents and nonagents. They do so because the distinction matters. It makes a difference if the object we glimpse in the reeds at dusk is a log or a crocodile. Human infants learn to distinguish between the quick and the dead using clues about how things move and the noises they make (dogs move differently from cars and make different kinds of noises), and how they interact with other objects.[48] But the neurological machinery that draws these distinctions is far from perfect, so it is not always easy to distinguish between what today we call the natural and supernatural realms.[49] Our brains are always on the lookout for agents, and it is so easy to make mistakes when we hear whispers behind us in the night. Why do iron filings creep toward magnets? Why do rivers in flood seem so angry? Dreams and hallucinations encourage us to believe in the possibility of many types of purposeful beings. So does language, because grammatical forms tell us that actions require actors. In English, grammar forces you to say that the wind blows, the sun shines, the world spins, the pandemic spreads. Our minds have a bias toward overdetection of agency because that is usually a less dangerous error than the alternative.[50] Mistaking a log for a crocodile might cause some merriment, but mistaking a crocodile for a log could prove fatal.

In short, the almost universal belief in a realm of purposeful beings and forces — the idea known to nineteenth-century anthropologists as *animism* — may originate from the way our minds work. That may be why most human communities have regarded the presence of spirits as self-evident. Even in the skeptical mind of Cicero, the spirit world was an empirical matter, so that, as his biographer Elizabeth Rawson puts it, divination could be regarded as "a branch of 'physics,' or study of the natural world."[51]

Elizabeth Marshall's account of the spiritual world of the Ju/wasi in the 1950s offers a glimpse into the spiritual beliefs that may have shaped many aspects of future thinking and management in the foundational era. The Ju/wasi knew many gods, even creator gods, but they thought of them as hunter-gatherers. Though powerful, their gods lacked the grandeur and pride of the imperial gods of more recent world religions. They looked "like human beings of normal, human size. They hunted, just as people did, and had wives and children and lived in camps with fires and grass shelters, just as people did."[52] The gods were neither mentors nor teachers of humans, though they could be unpredictably dangerous. But they could also be foolish.

For the Ju/wasi, as for many small-scale communities, contacting the world of spirits through ritual was an important way of dealing with uncertain futures, particularly in matters concerning health. Elizabeth Marshall witnessed trance dances that began at dusk, as the sun set and the full moon rose in the east.[53] Women would start a fire near the main encampment. Then, one by one, they would sit down on their heels near the fire. As stars began to come out, some would start singing and clapping and others would gradually join in, creating complex contrapuntal rhythms and songs of "heart-stopping beauty." The men, some with rattles around their legs, would start dancing in a circle around the women. Eventually, some of the men would fall into trances and start "washing themselves" in the flames of the fire. Then they would approach one of the women, place a hand on the woman's chest and back, and, suddenly standing up with a scream, seem to draw something out of the woman, a sickness, which they then threw to the spirits of the dead.

The dancing usually continued until dawn, when it would reach a culmination before ending suddenly, as the sun rose. "The tired women would stand up stiffly, having been sitting on their heels for almost twelve hours. . . . They would talk, laugh, stretch, and look around for leftover firewood." Ancient rock art from the Kalahari, some of which may be several thousand years old, suggests that similar traditions and worldviews may have roots that reach deep into the foundational era.[54]

CHAPTER 6

Future Thinking in the Agrarian Era

I devised the many methods of divination (mantikē), and I first judged what truth there is in dreams, and I first made known to mortals the meaning of chance utterances, hard to interpret, and of the omens one encounters while on the road; and I defined the flight of crooked-clawed birds — I explained which of them were auspicious or inauspicious by nature, . . . and I also taught mortals about the smoothness of entrails and what color the gall ought to have in order to please the gods, and all about the dappled beauty of the lobe of the liver. It was I who burned thigh-bones wrapped in fat and the long shank bone.

— PROMETHEUS, FROM AESCHYLUS'S *PROMETHEUS BOUND*[1]

The Agrarian Era of Human History

The agrarian era of human history began about ten thousand years ago (from about 8000 BCE). It ended about two centuries ago as the new technologies and ways of thinking of the fossil-fuels era began to lay the foundations for today's world.

After several hundred thousand years of extremely slow change in human societies, the transformations of the agrarian era were spectacular.[2] That era accounts for less than one-twentieth of the time since humans first evolved. But during that period, agriculture and other new technologies, driven by the accelerating pace and scale of collective learning, revolutionized human societies and ways of thinking. The unusual climatic stability of the entire era since the end of the last ice age (the period geologists call the Holocene epoch, which began about 11,500 years ago), allowed agriculture to spread through the world, laying the technological and demographic foundations for the even more profound changes of the modern era.[3]

Agriculture could produce much more food from a given area than hunting and gathering, and the surplus allowed human populations to grow from perhaps six million at the end of the last ice age to about nine hundred million by 1800 CE, at an average growth rate of almost 0.05 percent per annum. By 1800, most people lived not in mobile camps but in the sedentary communities we call villages, and about 7 percent lived in cities. The largest settlements had grown from fewer than a hundred people in the foundational era to more than a million. Over the same period, total human consumption of energy increased from fifteen million gigajoules per year to more than twenty thousand million gigajoules per year, while energy consumption per person rose by more than seven times, from about three gigajoules per year to about twenty-three.

Meanwhile, the three major trends powered by collective learning gathered pace. New technologies enhanced human control over the environment; larger networks of exchange synergized collective learning and increased the relative importance of social time; and the accelerating pace of change made time seem more dynamic and the future less predictable.

The breakthrough technologies of agriculture intensified human manipulation of the environment and made for a more manipulative relationship to the world and the future. Farmers found they could dramatically increase future yields by rearranging landscapes and altering the animals and plants around them through domestication. Some traditions even taught that the high gods had granted humans power over all other species. After destroying most life on Earth in a worldwide flood, the God of the Jews instructed the survivors, Noah and his family, that henceforth "the fear of you and the dread of you shall be upon every beast of the earth, and upon every fowl of the air, and upon all that moveth upon the earth, and upon all the fishes of the sea; into your hand are they delivered."[4] Agriculture became obligatory — if your community didn't farm, you could be sure that sometime in the future your farming neighbors, with more people and more resources, would squeeze you out. Competition between agricultural communities stimulated other new technologies, from pottery and metallurgy to new ways of building and new forms of transportation and communication.

New transportation technologies such as sailing, horse riding, and the use of oxen and camels widened human exchange networks, while new communication technologies such as writing multiplied links between communities and between generations. More information collected in expanding webs of exchange, and that catalyzed innovation. Time itself shape-shifted, as people found they had to march to the social rhythms of millions of other people within evolving grids of trade, ritual, warfare, and governance. Even in the most remote farming villages, the marketing of produce and the payment of taxes forced households to coordinate their activities with those of distant towns and rulers.

The pace of change accelerated, eroding belief in a stable universe. Social relations were revolutionized by the creation of the first cities and states from about five thousand years ago. Huge, hierarchical societies appeared, dominated by small, powerful, wealthy elites. The creation of states was a political innovation of vast significance because states, by their very nature, tried to manage futures on very large scales. Writing and the building of durable public monuments such as pyramids and palaces heightened awareness of change by preserving evidence of events in the distant past. Writing evolved to help elites keep track of their assets — their sheep, their slaves, their hoards of gold. But soon writing became essential for future planning in general because written documents could hold more knowledge in more stable forms than human memories, and that made it easier to track trends from the deep past. One of the oldest of all written stories, the *Epic of Gilgamesh*, tells us that its hero brings "information of (the time) before the Flood."[5]

Elite and Popular Future Thinking

In the agrarian era, as in all eras of human history, much, perhaps most, future thinking was commonsensical, trend based, intuitive, and empirical, and practical expertise was respected. "Do you think," asked a skeptical Cicero two thousand years ago, "that a prophet will 'conjecture' better whether a storm is at hand than a pilot? Or that he will by 'conjecture' make a more accurate diagnosis than a physician, or conduct a war with more skill than a general?" Though everyone used divination — attempts to contact and negotiate with beings from the spirit world — Cicero insisted that "divination is not applicable in any case where knowledge is gained through the senses."[6]

Even in the agrarian era, much, perhaps most future thinking was based on "knowledge gained through the senses." But in this chapter, we will focus on those aspects of agrarian-era future thinking that are less familiar today. Most of these arose from the assumption, taken for granted by Prince Arjuna and by most people before modern times, that beings or forces from the spirit world know of our futures and can shape them. "Come up hither," says a voice "as it were of a trumpet" to St. John of Patmos, "and I will shew thee things which must be hereafter."[7] Divination took many different forms. (Thomas Hobbes, who despised most forms of divination except those based on Protestant Christianity, gives a wonderful list in *The Leviathan*.)[8] And almost everyone respected divination, so few took seriously Cicero's careful distinctions between empirical and divinatory knowledge, and most respected the expertise of prophets and diviners as much as that of doctors, pilots, and generals.

Everyone used divination. But in societies shaped increasingly by divisions in class, power, culture, and wealth, the future thinking of educated elites began to diverge in important ways from that of most people. In remote villages and working-class urban tenements, people worried about their personal futures, assessed local trends, consulted local gods, spirits, and witches, and trusted in local divinatory traditions. But those in positions of power had to think about the futures of hundreds of thousands or millions of people. That meant looking beyond local traditions for larger and deeper trends and more authoritative spiritual voices. Not just: "Will locusts eat our barley field because our neighbors hexed us?" but: "Will locusts create an empire-wide famine, and can I prevent that from happening?" Not just: "Will I get sick because my cousin cursed me?" but: "Will plague kill off most of my subjects, and what can I do about it?" Future thinking on large scales required new types of knowledge and new ways of thinking about the future based on large trends that might shape millions of lives. And that encouraged the sort of questioning of traditional forms of future thinking that we find in Cicero's writings on divination. Elite future thinking also became more ambitious during this period because emperors and kings could shape the futures of millions by sending armies to distant lands or building new cities or diverting rivers or clearing forests. Finally, the elaborate rituals and belief systems that surrounded elite future thinking gave it an authority and prestige that most popular future thinking lacked.

Some distinctive features of elite future thinking are captured in the historical literature on what Karl Jaspers called the

"Axial Age," an era of significant political and cultural change in the first millennium BCE.[9] As networks of exchange expanded, there appeared the first trans-Eurasian networks of trade, and the first empires large enough to touch the edges of the known universe, so that their leaders could think of themselves and their gods as universal rulers. These empires included the Persian Achaemenid Empire, founded in the mid-sixth-century BCE, and the first unified Chinese empire, founded late in the third century BCE. The rulers of these vast and diverse empires tried to look beyond the local gods and traditions of the many peoples they ruled, in search of deeper and more universal trends and principles, whether laid down by a universal creator god or built into the fabric of reality. And in this way, argued Jaspers, there emerged the first religious and philosophical traditions that aspired, like modern philosophy and science, to universal, rather than merely local truths. As the historian Arnaldo Momigliano put it, one of the key features of the Axial Age was the search for "a more universal explanation of things."[10] We also see a widening "circle of concern" during this period, as universal religions, constructed within Eurasia-wide networks of power and exchange, began to generate religious and political identities and loyalties extending across "imagined communities" of millions of people.[11]

Most of the prophets and scholars who formulated the universalist worldviews of the Axial Age were literate, well-traveled or well-connected, and enjoyed elite patronage. The Persian prophet Zoroaster was one of the first to imagine a god whose laws embraced the entire universe, and the Achaemenid Empire adopted Zoroastrianism as its state religion. Claims for the existence of a universal order also appeared within the monotheistic traditions of Judaism, within the Indian Upanishads and in Buddhism, in the great Chinese philosophical schools of Confucius, Lao-Tzu, and others, and in the religious and philosophical systems of the ancient Greeks.

The universalism of Axial Age thinking was confined to the educated and powerful. And these elites were well aware of the differences between their thinking and that of most people. Priests and nobles and philosophers disdained the parochialism and what they saw as the bizarre superstitions of popular future thinking, though few were willing to abandon the realm of spirits and gods entirely, and even the most skeptical, such as Cicero, respected some forms of divination. After all, he argued, if we admit the existence of gods, how can we deny the possibility that they can communicate with us through divination? Nevertheless, thinkers like Cicero knew they thought differently from most people. "We Roman augurs," he wrote, "are not the sort who foretell the future by observing the flights of birds and other signs," even if, "out of respect for the opinion of the masses and because of the great service to the State we maintain the augural practices, discipline, religious rites and laws, as well as the authority of the augural college."[12]

Conflicts between Elite and Popular Future Thinking

Differences between popular and elite future thinking were most apparent when the two worlds clashed. That tended to happen when rulers tried to suppress forms of popular future thinking that contradicted or threatened their own authority as managers of the future.

In the pastoral-nomadic societies of the Eurasian steppes, empires rose and fell so fast that these conflicts could take spectacular forms. In the early thirteenth century, the Mongol followers of Genghis Khan lived in just one generation through the transition from small pastoralist clans with traditional religious practices and beliefs to a global empire reaching across much of Eurasia and influenced by the universal religions of the Axial Age. The shamans who dominated popular future thinking in the traditional Mongol world were powerful, and many were chiefs.[13] They healed the sick, went into trances, observed the heavens, and predicted eclipses. They used the cracks on burned sheep bones to make forecasts, identified witches, and nominated favorable days for making war or nomadizing, and some used special stones to manage the weather or whip up blizzards. Genghis Khan himself claimed shamanic powers. According to the Persian historian Juzjani, he was "an adept in magic and deception, and some of the devils were his friends. Every now and again he used to fall into a trance, and in that state of insensibility all sorts of things used to proceed from his tongue."[14]

As Genghis Khan (Temüjin) rose to supreme power, the most important shaman in his entourage was Teb Tenggeri (Kökechü), who was reputed to ride to heaven on a gray horse, to walk naked in the coldest of winters, and to turn water into steam. The two had been friends since youth. Teb Tenggeri announced that Heaven had chosen Temüjin as future lord of the world and may have first bestowed on him the title Genghis Khan, or "Universal Ruler."[15] But as Genghis Khan's power increased, he found himself employing and ruling people from many different cultural and religious traditions, including Buddhists, Taoists, and Muslims, and his own philosophical and religious horizons began to widen. Eventually, he acquired something of the universalist spirit of the Axial Age. The Muslim historian Juvaini, who was employed by Genghis Khan's successors, wrote that Genghis Khan "eschewed bigotry, and the preference of one faith to another and the placing of some above others; rather he honoured and respected the learned and pious of every sect, recognizing such conducts as the way to the Court of God."[16]

Genghis Khan's shifting perspectives may help explain the looming conflict between the two shamans. By 1210, Teb Tenggeri and his family were beginning to undermine Genghis Khan's authority. They threatened his brother Otçigin and other members of his family, lured away some of his followers, and Teb Tenggeri predicted that Genghis Khan might be losing the mandate of Heaven.[17] According to *The Secret History of the Mongols,* when Genghis Khan heard that Teb Tenggeri and his brothers planned to visit him, he said to his brother Otçigin, who had his own reasons for distrusting the shaman, "Teb Tenggeri is coming now. Whatever you may wish to do to him within your power, it is for you to decide." Otçigin waited with three "strong men" and, when Teb Tenggeri arrived, challenged him to wrestle. "Otçigin seized the collar of Teb Tenggeri, saying, 'Yesterday you compelled me to make amends. Let us now measure up to each other!'" Otçigin dragged Teb Tenggeri outside Genghis Khan's tent, where the three strong men were waiting. They broke the shaman's back and threw down his body — a form of execution used for those of high status to avoid the spilling of blood. On the third night after his death, Teb Tenggeri's body vanished from the tent in which it had been placed, proof, claimed Genghis Khan, that even the heavens had rejected him. As historian Christopher Atwood writes, "Chinggis Khan thus replaced Teb Tenggeri as the empire's voice of heaven's will."[18]

Many political and religious crosscurrents lay behind this battle. But in part it was a contest for authority over the future between a traditional shaman whose perspectives were local and personal, and a rising emperor with a broader and more universalist vision. That vision survived Genghis Khan's death. In 1254, the Christian ambassador William of Rubruck was present as Genghis Khan's grandson, Emperor Möngke, listened respectfully to advocates of different religions before announcing that "we Mongols believe there is but one God. . . . But just as God gave different fingers to the hand so has He given different ways to men."[19]

Though not always this violent, conflicts between different

visions of the world and the future were ubiquitous in the agrarian era.[20]

Elite Future Thinking in the Agrarian Era

We have a lot of written evidence on the future thinking of the educated and powerful. The sources are particularly rich for Greece, Rome, Mesopotamia, and China, and the following sections are based mainly on evidence from these regions.

Future Thinking in Classical Greece and Rome

The Mediterranean world two thousand years ago comprised many different types of communities and polities, from remote villages to colonizing city-states to vast empires, so it offers evidence of many different, overlapping, levels of future thinking.

Divination was universal. It was taken for granted, banal, a matter of common sense. In a recent study of divination in Greece, the classicist Sarah Johnston writes:

> It is likely that in antiquity, most people practiced or witnessed some form of divination at least once every few days: divination was always part of offering sacrifices to the gods, usually part of deciding whether to undertake a military maneuver, often part of puzzling out a bewildering dream, sometimes part of diagnosing and treating an illness or choosing a bride, and even, sometimes, part of understanding why your body was twitching or your child was sneezing. Walking through the ancient marketplace, you might glimpse a "belly-talker" who carried a prophetic spirit around inside of herself, an Orphic priest who could tell you what it meant if a weasel had

crossed your path, or a state delegation setting out to consult the Delphic Oracle on a matter of public good.[21]

The *Anabasis,* which describes Xenophon's long march through Persia, explains how divination could shape battle tactics and morale: "The *manteis* [diviners] were performing *sphagia* [animal sacrifices] into the river. The enemy were shooting arrows and slings but they were still out of range. But as soon as the *sphagia* were favorable all the soldiers began singing the paean and raising the war cry."[22] How could you not take divination seriously when the deepest thinkers of ancient Greece and Rome believed, and not just in the gods and spirits of official theology. Augustine believed that *daimones* carried messages between the human and spirit worlds. And as we've seen, even the skeptical Cicero defended some forms of divination. Indeed, the historian Mary Beard has argued that he took augury extremely seriously, though he saw it as a way of seeking divine approval rather than a way of peering into the future.[23]

Many Greek methods of divination borrowed from Babylonian and Assyrian traditions.[24] Like their Mesopotamian counterparts, Greek diviners watched the flight of birds as they ascended to the gods, they studied the organs of sacrificed animals, they listened to dreams, they noted strange events, they cast lots, and they interpreted omens.

Particularly solemn were the prophecies uttered by oracles in sites or temples, such as Delphi, that were managed by local priests and nobles. Consulting the Oracle of Delphi was a serious business, theatrical and awe-inspiring and costly in money and time. As you climbed up to Delphi, you knew you were near the dwelling of the gods. Power hummed around you. Delphi itself, on Mount Parnassus and with majestic views over the Gulf of Corinth, was enchanting. It was remote, so getting there took time, and when you arrived, you found a whole community serving you and the oracle. There were inns, hotels, hawkers, and shops where you could buy animals to sacrifice.[25] You paid to consult with the prophetesses, or "pythia," who transmitted messages from Apollo at special ceremonies that happened only a few times each year.

The divinatory rituals at Delphi included shamanic elements. The pythia spoke in trances thought to be induced by "vapors" present in the caves. Modern geological research has shown that ethylene, ethane, and methane do indeed leak from the caves at Delphi, and the sweetish smell of ethylene even matches contemporary descriptions from sources including Plutarch.[26] The obscure utterances of the pythia were translated by priests, so for all your expense and effort, you were always at least two degrees of separation from the great god Apollo. Thomas Hobbes, who scoffed at most forms of divination, insisted that the answers of ancient oracles "were made ambiguous by designe, to own the event both ways or absurd by the intoxicating vapour of the place, which is very frequent in sulphurous Cavernes." The futurist Oona Strathern argues that the real divinatory work was done by the priests, who "used their intelligence as well as a wide network of contacts, gossip and messengers to gather relevant information that would provide astute or 'useable' answers, thus maintaining customer satisfaction."[27]

We have plenty of evidence about the sorts of questions that were put to oracles because many divinatory institutions kept detailed records of questions and answers. Not surprisingly, most questions fall into the Red Zone of Future Cone 3, which maps Domains of Anxiety. Most were about individuals and families and their microfutures. Will we be healthy? Will we be happy? Why am I sick? Who made me sick? Will we have children, and will they prosper? Should I take this job? Is someone cheating me? The oracle of Dodona in Greece recorded questions and answers on lead tablets from the sixth to the third centuries BCE. They include the following:

- Geris asks Zeus concerning a wife, whether it is better for him to take one.
- Heracleidas asks Zeus and Dione . . . whether there will be any off-spring from his wife Aigle.
- Lysanias asks Zeus Naios and Deona whether the child with which Annyla is pregnant is not from him.
- Cleotas asks Zeus and Dione whether it is better and profitable for him to keep sheep.[28]

More official and broader in scope were the questions put to Greek oracles by city-states or their emissaries. In 426 BCE, a delegation from Sparta asked the oracle at Delphi whether they should establish a colony at Heracleia in Trachinia. They were advised to do so. In 432 or 431 BCE, a Spartan delegation asked whether they should attack Athens. According to Thucydides, the oracle replied, "If you go to war with all your might you will have victory and I, Apollo, will help you, both when you call for my aid and when you do not."[29] This last response is a reminder that even the renowned oracle at Delphi could offer answers as slippery, as open-ended, and as unhelpful as those in a modern fortune cookie. But it also reminds us that even the fuzziest of predictions can influence the future, because the Spartans did attack, launching the Peloponnesian War, which would last for almost thirty years. Would they have attacked if the oracle had given a different answer?

Future Thinking in the Bureaucratic Empires of Mesopotamia and China

At the upper levels of large bureaucratic empires such as those that emerged in the second and first millennia in Mesopotamia and China, we find approaches to future thinking that are more general and impersonal because they concern the fates not just of individuals or families but of whole societies. In a comparative study of Greek and Chinese techniques of divination, the historian Lisa Raphals notes that in the relatively small societies of classical Greece, most questions were addressed to particular gods, while imperial Chinese divinatory practices assume "a different, more mechanical, and arguably more naturalistic picture."[30] In the hands of powerful rulers and their high officials, future thinking was also more tightly controlled because of its political significance. Often, it shaded into propaganda.

Mesopotamian Traditions

Perhaps our oldest direct evidence of official divination comes from a collection of letters written in the Mesopotamian city of Mari in the eighteenth century BCE. These contain indirect accounts of prophecies that may have been massaged after the fact to support particular interpretations of the events they describe. Most describe messages from gods to rulers. For example, one letter reports messages from a prophet of the god Samas to the king of Mari, Zimri-Lim (ca. 1774–1760 BCE). One section reads:

> Thus says Samas: "Hammurabi, king of Kurda, has [talked d]eceitfully with you, and he is contriving a scheme. Your hand will [capture him] and in [his] land you will promu[lgate] an edict of restoration. Now, the land in [its entirety] is given to your hand. When you take con[trol] over the city and promulgate the edict of restoration, [it sho]ws that your kingship is etern[al]."[31]

This looks like official divination in its simplest form: a powerful mortal seeking advice from the gods. But it may also be a form of propaganda, a way of reminding citizens and enemies that Zimri-Lim has powerful divine allies.

A thousand years later, official diviners in Assyria kept elaborate records in the seventh-century library of Ashurbanipal in Nineveh, which contained more than three hundred clay tablets, the equivalent of thousands of pages of modern printed text.[32] In these records, the voices of the gods are more muted and robotic. Divinatory accounts from Nineveh involve less direct contact with the gods and more interpretation of signs, which suggests a more empirical and impersonal approach to future thinking. For example, they offer technical and mechanical explanations of how to interpret the entrails of sacrificed animals. When studying the liver of a sacrificed sheep: "If the base of the Presence is long and descends to the *right* Seat of the Path: The enemy will carry off the land of the prince, in battle the enemy will rout me and stand in my camp." On the other hand, "If the base of the Presence is long and descends to the *left* Seat of the Path: The prince will carry off the land of his enemy, in battle I will rout the enemy and stand in his camp." These tablets may have been used to train official diviners, like the many clay or bronze models of animal livers that have been found in the Near East.[33]

It is not clear whether the signs read in animal entrails were interpreted by seventh-century BCE Assyrian diviners as direct messages from the gods or as impersonal hints of cosmological trends and regularities. At the time, such regularities were studied with particular care in the heavens, as Mesopotamian astronomy and astrology flourished in the first millennium BCE. Originally seen as a way of determining auspicious or inauspicious times for activities such as launching wars, the study of the heavens would eventually merge with the Axial Age intuition that cosmological laws pointed to universal regularities and trends reflecting the will of imperial gods or perhaps simply the impersonal rationality of the universe.[34]

Chinese Traditions

In China the earliest evidence of divination comes from the era of the Shang king Wu Ding (ca. 1200–1181 BCE). As in Mesopotamia, early examples of divination include direct communication with the gods, particularly ancestor gods. But even the earliest Chinese records lack the ecstatic or trancelike quality that we saw in the Delphic oracle. Their tone is sober, empirical, and impersonal. Shang diviners addressed ancestral spirits as if they were government officials, each with their own rank, office, and expertise. Minor questions might be addressed to minor ancestors. But the big questions about war and peace or the harvest were addressed to the most senior ancestors, or to the high god, Di, the spiritual counterpart of the emperor, and the only god who could control the wind and the rain.[35]

Since the late nineteenth century, Chinese archaeologists have found huge collections of oracle bones, mostly cattle scapulae or turtle shells. They first turned up in 1898 in villages near Anyang in north Henan Province, where they were sold as "dragon bones." Scholars soon realized that the marks they bore were the earliest examples of Chinese writing. Since then, some two hundred thousand oracle-bone inscriptions have been found in China, about fifty thousand of which have been published.[36]

Divination using scorch marks on bones is known as scapulimancy. The practice is widespread, but it became particularly important and formalized in official Chinese divination from the late second millennium BCE. Shang kings or their diviners carved their questions into turtle plastrons (the belly side of turtle shells) or the scapulae (shoulder bones) of cattle, and answers appeared in the shape of cracks that formed when the bones were heated. Most questions touched on high policy, so divination was a serious matter, which is why Shang kings invested money, personnel, and time in careful divination.[37] Over time, divinatory bureaucracies became increasingly elaborate. Zhou kings of the third century BCE had three divinatory officials, each with his own staff: a director of divination, a director of incantation, and a director of astronomy. The first did the divining (using turtle shells, dreams, and the casting of milfoil [yarrow] stalks), the second called up spirits, and the third recorded the results. The astronomer, or "Taishi" (a role held later by the famous Han historian Sima Qian), also handled calendrical calculations and identified auspicious days for events and decisions.[38]

Shang kings collected vast numbers of cattle bones and turtle shells, often in the form of tribute. They were cleaned, prepared, and ritually consecrated before questions and responses were carved into them, along with the names of the diviners. After use, the bones were stored in special archives.[39] The following account by David Keightley, an expert on Chinese divination, suggests how meticulously imperial divinatory rituals were choreographed.

> Hollows were chiselled or bored into the backs of scapulas and plastrons with great care and effort so that the pu-cracks, which appeared on the front when the diviner burned the hollows, formed in their turn a series of preordained patterns. Unlike the free-form pyromancy of other cultures, where the bone might be thrown into the fire, or the heat applied to any point of the unprepared bone surface, there was nothing random about the pyromantic cracks of the Late Shang. . . . No crack could appear where the Shang diviner did not want it to. The powers could not reveal themselves in unexpected ways. The supernatural responses were rigorously channelled.[40]

The range of possible answers was limited, not just by the way the bones were prepared, but also by the questions, which

Figure 6.1: Oracle Bone (Scapula) Recording Several Divinations from the Reign of the Shang king Wu Ding (From Keightley, "The Shang," 243)

were posed laconically, in the form of simple alternatives, for example:

"There will/will not be, a good harvest"
"It will rain/it will not rain"
"The king should ally (or not ally) with this tribe"
"The king should attack (or not attack) that one"
"Fu Hao's childbearing will be good/not good"

Official diviners had a lot of control over predictions. That suggests that many questions were not really requests for *information* about the future but attempts to *manage* or *demonstrate control over* the future. As David Keightley argues, some divinatory rituals were really "spells applied to the future." In this view, carving into a bone or shell that the harvest will be good is a way of managing the future by analogical magic.[41] Influencing the harvest was, of course, a vital matter because good harvests meant larger revenues, future prosperity for peasants, and divine blessing on their rulers. The theatrical and propagandistic aspect of official divination is particularly clear in what Keightley calls "display inscriptions." These listed royal forecasts and subsequent evidence of their accuracy. They show rulers advertising how much power they wielded over the future, so they are about legitimation as much as prophecy.[42]

The questions posed by Chinese kings and emperors were about the weather and the harvest, about whether projects and plans would succeed, about whom to appoint to important positions, about the meaning of strange events or dreams. There were also personal questions about the royal family, marriages, births, royal succession, and the health of the ruler. But in an imperial context these questions were never *just* personal, because the answers had great political significance.[43] Many questions took the general form: Will my actions turn out well? Is this the right time to act? By the Han era, beginning around the end of the third century BCE, daybooks were widely available, with lists of auspicious and inauspicious times for different activities. Many offered travel advice:

Returning Home: In all cases in the third month of spring on a *ji* or a *chou* day you cannot go east. In the third month of summer on a *wu* or a *chen* day you cannot go south. In the third month of autumn on a *ji* or a *wei* day you cannot go west. In the third month of winter on a *wu* or a *xu* day you cannot go north. [Travel] within a hundred [*li*] is severely inauspicious. Travel beyond two hundred *li* is fatal.[44]

In the first millennium BCE, during Jaspers's "Axial Age," official Chinese divination became less concerned with the advice of ancestors and more interested in deep cosmological trends and regularities. As historian Lisa Raphals puts it, "Chinese methods became increasingly independent of direct human-divine interactions. The systematic outlook that informed most Chinese divination methods by the late Warring States [fifth to third centuries BCE] was entirely compatible with naturalistic inquiry."[45] David Keightley interprets these shifts as signs of increasing confidence that both heaven and earth were subject to universal cosmological laws. Chinese official religion and divination acquired a "this-worldliness" that would become characteristic of Chinese philosophy in general.[46]

The philosophical systems of Confucius and Daoism, which took shape during the middle of the first millennium BCE, were concerned mainly with universal principles of ethics and existence. Symptomatic of these shifts toward a more impersonal cosmology is the increasing importance of astronomy. Astronomy occupied an ontological borderland where the will of the gods and more impersonal laws and forces contended for control over the future. Were unexpected astronomical phenomena such as comets or new stars (supernovas?) to be understood as signs from the gods or evidence of the working of an impersonal cosmological machinery? Over ancient astronomy, in China and in many other agrarian civilizations, there always hovered the intuition that the heavens can influence our lives quite independently of the gods.[47] That belief encouraged fatalism as well as forecasting. As the villain, Edmund, says in Shakespeare's *King Lear:* "This is the excellent foppery of the world: that when we are sick in fortune — often the surfeit of our own behavior — we make guilty of our disasters the sun, the moon, and the stars, as if we were villains by necessity, fools by heavenly compulsion, knaves, thieves, and treacherers by spherical predominance."[48]

Chinese astronomer-diviners used astrolabes with two main parts: a circular "Heaven" plate, which could be set to the present day and time, and a fixed "Earth" plate aligned to the four cardinal directions.[49] A third-century BCE text, *The Rites of Zhou*, explained how the royal astronomer tracked the movements of heavenly bodies and clusters of stars "in order to discern [corresponding] trends in the terrestrial world, with the object of distinguishing (prognosticating) good and bad fortune." Different regions of the empire depended on different celestial bodies, whose motions predicted their "prosperity or misfortune." Unlike Western astronomy, that of China focused on the northern sky and treated Ursa Major (the Big Dipper) as a sort of celestial clock hand.[50]

The following example, from Sima Qian, suggests how astronomical observations could be used in official divination. King Yuan of Song had a disturbing dream and called his diviner, Wei Ping, to explain it.

> Wei Ping stood up, adjusting the mantic astrolabe with his hands [presumably to set them to the right date and hour]. Raising his eyes to heaven he gazed at the light of the moon; he looked to see where the Dipper was pointing and determined the position where the sun was situated. As aids he used a compass and square along with a weight and scales. When the four nodal points were fixed and the eight trigrams were face to face, he looked for the signs of good or ill fortune.[51]

The history of the *I Ching* illustrates the slow retreat of gods and spirits from elite divination, and the increasing importance of more impersonal and philosophical modes of future thinking.[52] The most ancient form of the *I Ching* is the *Zhou I*, a collection of divination formulas that probably dates to early in the Zhou era (1050–771 BCE). Its formulas were linked to sixty-four different "hexagrams," each consisting of two "trigrams" made of three lines that could be either continuous or divided in two. Over time, commentaries and interpretations gathered like moss around the original formulas and hexagrams to create an extremely rich, complex, and often abstruse system of divination. By the Han era, such commentaries had been formalized in the "Ten Wings," which offered standard explications of the hexagrams. The rich interpretative literature that gathered around the hexagrams became deeply embedded within Chinese thought and philosophy and helped formalize and enrich the ancient cosmological binary of yin (broken lines) and yang (solid lines).

It may be appropriate to think of the hexagrams, along with the interpretative literature that accreted around them, as a sort of encyclopedia of life situations and possible futures, written in the opaque language typical of prophetic utterances. To divine using the *I Ching*, it was necessary to pick a hexagram randomly. Often this was done by throwing milfoil (yarrow) stalks to generate numbers from which the hexagram could be built up, from the lowest to the highest line. That was the easy part.

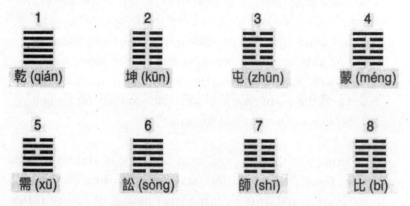

Figure 6.2: The First Eight of the Sixty-Four Hexagrams of the *I Ching* (From the Wikipedia article on the I Ching: Wikipedia, s.v. "I Ching," last modified October 29, 2021, 20:00 (UTC), https://en.wikipedia.org /wiki/I_Ching)

Interpreting the gnomic "judgments" associated with each hexagram was so complex a task that Karl Jung, who was fascinated by the *I Ching,* once said, "The less one thinks about the theory of the *I Ching,* the more soundly one sleeps."[53]

The first hexagram, "Qian," or "Heaven," consists of six unbroken lines and represents "The Creative," while the second, "Kun," or "The Earth," consists of six broken lines and represents "The Receptive." The "judgment" associated with the first hexagram contains a number of statements: "Begin with an offering; beneficial to divine; Hidden dragon, do not act; The dragon is seen in the field; it is favorable to see the powerful person; The upright person strives energetically all day long, Vigilant day and night, thus averting blame; Sudden leap into the whirlpool, blame averted," and so on.[54] The meaning of such formulas was obscure even for contemporaries. But as with many forms of divination, the obscurity was part of the process. On the one hand, it made it harder to disprove a forecast. But it also forced clients to think carefully about possible hidden meanings and make use of their own intuitions as well as the many available commentaries, because in practice divination was not just about forecasting; it also helped open people's minds to new possibilities.

The shift we have described in Chinese official divination from inspirational to more impersonal forms of future thinking was subtle, and its impact should not be exaggerated. Sacrificial rituals remained ubiquitous and fundamental even in elite circles, and sacrificial offerings always implied a relationship with *beings* rather than mere *forces.*[55] But at lower levels of Chinese society, we can be sure that divination was less austere and more personal. Diviners known as *fangshi,* who could be hired by anyone with the right cash or connections, used magical powers to heal from a distance and revive the dead. They could predict eclipses and even the time of their own death. More inspirational in their techniques were the practitioners known as *wu,* who, like

Siberian shamans, practiced magic and exorcism, controlled the weather, and talked with spirits. But even they were sometimes included in official rituals. "They performed dances to bring rain during drought; preceded the queen in visits of condolence . . . and sang, wailed, and prayed during great calamities of the state."[56] That such inspired divinatory activities could be found at the highest levels of Chinese society makes it certain that in China's villages and towns, more traditional forms of divination flourished.

Glimpses of Popular Future Thinking

We have much less evidence on the future thinking of most people in the agrarian era. They surely shared many of the ideas, methods, and rituals we have described so far because the cultural membranes between classes were thin enough for ideas and attitudes to leak across them, particularly in villages or rural estates where masters, workers, servants, and slaves were in daily contact. Nevertheless, even the most superstitious among the educated elite knew they lived in a different intellectual world from that of most people.

To get a glimpse of popular forms of future thinking and divination we have to turn once again to the Nasreddin Hoca strategy. Here, that means relying on modern accounts of popular future thinking in communities that seem to have preserved traditional ways of thinking even as they have adapted to the modern world. As the anthropologist Ana Mariella Bacigalupo argues in a recent study of modern shamans and witches in southern Chile: "The roots of these rituals are ancient, but *machi* [shamanic healers] today thrive as they engage with contemporary concerns and incorporate into their spiritual practices the knowledge and symbols of Catholicism and the national medical and political systems, transforming and resignifying them in the process."[57] Research of this kind shows the presence of a realm of popular future thinking shaped by spiritual beings and forces of many different kinds, who had to be contacted, negotiated with, and even fought in order to cope with the future. We see little trace of the search for impersonal, universal principles that was increasingly evident in elite future thinking in the agrarian era.

Though we find common features, the details of popular future thinking were extraordinarily diverse because they were shaped by local traditions. In Russian villages in the early twentieth century, atheistic Soviet officials were embarrassed to find approaches to future thinking that may have changed little for centuries. As the historian Moshe Lewin writes:

> All the many needs of a peasant household were taken care of by magic rites, formulas, potions, herbs, all helping to find the thief and recover stolen goods, to assure a successful birth (including that of domestic animals), to protect the newlywed (and everyone else) from the "evil eye," to preserve the family from ominous influences that a corpse in the house awaiting funeral could create. All the stages of the cycles of nature and of life require protection.[58]

For Russian peasants, preparing for the future was a never-ending nerve-racking game played with unpredictable, dangerous, and mostly invisible spirits and forces. Every household kept in touch with dead ancestors, who lived nearby and looked after their descendants. *Domoviki*, or household spirits, lived in the house. Other spirits tended the yard or protected the village bathhouse, a place full of magic, particularly after midnight.[59] Many spirits were dangerous and to be avoided. Exceptionally spooky spirits lived in lakes and rivers. They included the *rusalki*, or water nymphs, who were dead-but-alive (you could tell because their eyes didn't move) and would lure the incautious to terrible deaths. There were half-dead or undead spirits, including unbaptized infants and suicides. Mini-devils gathered at crossroads or in dark places. Some had tails. Some had their own families, households, and retinues. They were dangerous like their lord, the devil. They could snatch babies or make you deadly sick. But if you were cunning, you could sometimes bribe or trick them. To confuse wood sprites you could wear your clothes inside out or back to front.[60]

There were many domestic forms of divination in Russian villages. People asked village diviners how long they would live, how good the local harvest would be, and how to find thieves. Particularly popular were ways of seeing your future spouse, because marriage was the major event in most peasants' lives. The Soviet Union's long-serving foreign minister Andrei Gromyko described in his memoirs how young people in his home village tried to see their future spouse by taking a mirror and a torch to the bathhouse and waiting until midnight for an image to appear. In some villages, unmarried women placed a mirror, a bowl of water, and some barley grains on the floor and put a chicken next to them. If the chicken looked into the mirror, your husband would be a dandy; if it drank the water he was going to be a drunkard; if it pecked the grains he would be rich.[61]

Anthropologists have found similar practices in many parts of the world. The English anthropologist Edward Evans-Pritchard, who lived among the Azande people of the upper Nile in the 1920s, described the widespread use of "poison oracles." After asking your question, you administered a measured dose of a special poison to a chicken, and the death or survival of the chicken gave you your answer. Such methods, like many divinational technologies, gave diviners considerable power over divinatory outcomes. Here is an example from a session attended by Evans-Pritchard: "X's mother lies seriously ill. Is her sickness due to Basa? If so, poison oracle kill the fowl. If Basa is not responsible, poison oracle spare the fowl. The fowl SUR-VIVES, giving the answer 'No.'" A further question was then posed: "If the evil influence that threatens his wife is due to Mekana's household, then poison oracle kill the fowl. If the evil influence emanates from the wives of his wife's grandfather, then poison oracle spare the fowl. The fowl SURVIVES, con-firming that evil influence is from the homestead of the girl's grandfather."[62] Further questions were asked to narrow down the source of trouble and find what to do, and often questions were repeated for confirmation.

If traditional methods failed, or seemed inadequate, you could turn to specialists, people known for their skill in medi-cine or exorcism or forecasting or negotiating with the spirit world. Most villages had healers or people with a reputation for predicting or fixing the future. In 1925, a newspaper report from Tver Province in Russia described a local healer (*znakharka,* "one who knows") named Anisia Ivanovna, who was also known to be skilled in witchcraft and exorcism:

> If a husband quarrelled with his wife, if a cow failed to conceive, if a person or an animal fell ill, or if a young man broke up with his girlfriend, people would turn to "Mother Anisiushka" for help. Before they even entered her house, she would greet them with: "You are pos-sessed of the devil! Quickly, say a prayer!" She would raise her skirt over her head, climb up on the oven, or crawl under a table, emerging once the visitor had recited a litany of prayers. Only then did she inquire about the reason for the visit. She would make the visitor drink a cloudy brew that was supposed to exorcise the devil, or she would give him a potion to mix in his tea or feed to his infertile cow. The villagers persisted stub-bornly in the belief that there was something "holy"

about Anisia, and they would travel fifteen to twenty kilometres to seek out her assistance.[63]

It is tempting to think that the cosmological world of Anisia Ivanovna lurks just below much of the documentary evidence on popular religion in the agrarian era, and similar figures hover at the edge of all our evidence on ancient future thinking.

If problems were serious enough, you could turn to the professionals, who made a living by negotiating with the spirit world. In his autobiography, the Russian dissident priest Avvakum, who was transported to Siberia in the 1660s, described how one of the officers guarding the convoy forced a local Tunguz wizard to *shamanit'*, or tell their fortune:

> In the evening that wizard . . . brought a live sheep and began to work magic over it: he rolled it to and fro for a long space and then he twisted its head and flung it away. Then he began to jump and dance, and invoke devils, and, giving great screams the while, he flung himself upon the earth, and foamed at the mouth; the devils were pressing him, and he asked them: "Will the expedition prosper?" And the devils said: "It will return with much booty, having gained a great victory."[64]

Avvakum saw the shaman as an agent of the devil, prayed for the destruction of the prisoner convoy of which he was a part, and was delighted when, indeed, most members of the expedition were killed.

Since Avvakum wrote, the word *shamanism* has entered scholarly literature to describe a type of divination, often known as *trance divination,* in which specialists contact spirits while in a trance.[65] Descriptions of trance divination can be found in many different parts of the world. Most practitioners were highly trained, while some inherited their skills and some were called to the role, often against their will, or during great personal crises. Ana Bacigalupo recorded how a modern *machi*, or shaman, from southern Chile remembered being initiated by spirits:

> Eeeeeeeh! I am going to be a *machi* [shaman] . . . that heals with herbal remedies, they told me. Many different types of medicinal plants came together. They frothed and flowered. Suddenly they gave me my instruments that would accompany me. "You will pass through all places on earth. You will ride on a horse. You will go everywhere," they said.[66]

Some educated commentators in the agrarian era believed that shamanic or divinatory talent was a developed form of skills that were dormant in most people but could be tapped in dreams. In Cicero's dialogue *On Divination,* his brother Quintus insists that the gods have granted everyone the power of divination, but in some it is abnormally developed, and is called "'frenzy' or 'inspiration,' which occurs when the soul withdraws itself from the body and is violently stimulated by a divine impulse." Most people lack these powers except when the soul is "so unrestrained and free that it has absolutely no association with the body, as happens in the case of frenzy and of dreams." Even Socrates preferred to trust those who divined while in trance, or "enthusiastic madness" as he called it, on the grounds that they were probably in direct contact with the gods.[67]

Shamans entered the world of spirits through trance, using methods that included dancing, drumming, and drug taking. They wore special costumes so that their spiritual helpers would recognize them, and once in the spirit world, they used the reciprocal courtesies and methods of negotiation familiar in all face-to-face communities. Some fought magical battles on behalf of their human clients. Piers Vitebsky describes a battle of revenge between Tunguz clans, which began when the shaman of one clan sent a worm to kill a member of another clan.[68] The worm snuck past the reindeer spirit guards of the neighboring clan and buried itself in the guts of its victim. A shaman from that tribe sent the spirit of a goose and a snipe to extract the worm, and an owl spirit to dispose of it safely in the underworld. Shamans also bargained for fertility, for success in battle, and for freedom from disease. In return, they offered gifts and sacrifices to those in the spirit world, or threatened them, or hectored or pleaded with them.

Evans-Pritchard described trance divination as practiced among the Azande. In one séance that he witnessed, a farmer asked whether this year's crop of eleusine (finger millet) would succeed. The answers contain a warning not about dangerous spirits, but about family members who might use magic to harm the farmer's crop. This meant that the diviner adjusted the question.

> The witch-doctor . . . dances because it is in the dance that medicines of the witchdoctors work and cause them to see hidden things. It stirs up and makes active the medicines within them, so that when they are asked a question they will always dance it rather than ponder it to find the answer. He concludes his dance, silences the drums, and walks over to where his interlocutor sits. "You ask me about your eleusine, whether it will succeed this year; where have you planted it?" "Sir," he replies, "I have planted it beyond the little stream Bagomoro." The witch-doctor soliloquizes. "You have planted it beyond the little stream Bagomoro, hm! hm! How many wives have you got?" "Three." "I see witchcraft ahead, witch-craft ahead, witchcraft ahead: be cautious, for your wives are going to bewitch your eleusine crop. The chief wife, it is not she, eh! No it is not the chief wife . . . not the chief wife, not the chief wife. Do you hear it? Not the

chief wife." The witch-doctor is now entering into a trance-like condition and has difficulty in speaking, save in single words and clipped sentences. " . . . Malice. Malice. Malice. The other two wives are jealous of her. . . . Do you hear? Jealousy is a bad thing. Jealousy is a bad thing, it is hunger. Your eleusine crop will fail. You will be troubled by hunger; you hear what I say, hunger?"[69]

This dramatic account is a reminder that the authority of divination was often heightened by performance skills, by elaborate costumes and sound effects and deliberately obscure utterances. Azande witch doctors wore costumes that included straw hats with bird feathers, wooden whistles, animal skins, rattles, and ankle bells, so that as they danced, each was, in Evans-Pritchard's words, "a complete orchestra."[70] Evans-Pritchard found that professional witches prepared for séances by tuning in to local gossip. Who was quarreling with whom? Who was seen slipping into whose bedroom? Knowing who might bear a grudge against the organizer of a séance made it easier to identify possible enemies who might have used witchcraft against them.

Once, Evans-Pritchard caught two witch doctors faking the removal of objects from a body during "surgery," but while they admitted the trickery, they insisted that their medicines worked and that was what mattered.[71] Most professional diviners probably saw such tricks as a legitimate part of their craft. But there were surely many out-and-out frauds as well. The second-century BCE Roman poet Ennius wrote of "village mountebanks, astrologers who haunt the circus grounds . . . or dream interpreters," and insisted, "they are not diviners either by knowledge or skill, — [b]ut superstitious bards, soothsaying quacks."[72] Fraudulent forecasters survived in the ancient world, as today, partly because of a principle that St. Augustine understood: if you make enough predictions you will get it right some of the time.[73]

Educated or not, most people understood that not all divin-ers were to be trusted. But that did not undermine people's gen-eral belief in divination. Trusting predictions often made sense because even the least competent diviners tried to make plausi-ble predictions, and almost everyone in the agrarian era believed in the ubiquity of magic and of spirit beings and forces. Besides, those who appealed to diviners or engaged in divina-tion were usually driven by deep anxieties that encouraged them to clutch at any answers that seemed authoritative.

Nevertheless, belief in divination did not rule out some degree of skepticism. Many Azande understood that diviners could be incompetent or fraudulent, but even when frauds were exposed, people continued to believe in witchcraft. As Evans-Pritchard notes, Azande witches were trusted in the same cau-tious way that modern doctors are trusted — not because they were always honest or successful, but because they had a lengthy training, and their prescriptions worked quite well quite often. "If one witch-doctor fails to cure an Azande he goes to another in the same way as we go to another doctor if we are dissatisfied with the treatment of the first one whom we have consulted."[74]

Many aspects of popular future thinking in the agrarian era may seem bizarre and naive to modern observers. But it is important to remember the profound anxieties that drove people to divination; the extraordinary precariousness and insecurity of most people's lives; the absence of many of the technological, medical, and legal protections of the modern world; the universal belief in a world full of spirits; and, finally, the sheer inexplicability of many of the threats and dangers that people faced in their everyday lives in a world without modern forms of science. In its time and place, popular future thinking offered powerful and credible forms of consolation and some hope of empowerment to the vulnerable, and it still does today for many who find themselves in the Red Zone of the Future Cone of Domains of Anxiety.

The Experience of Divination:
The Oracles of Astrampsychus

We will end this chapter by looking at a particular divinatory process described in a text known today as the "Oracles of Astrampsychus," which Mary Beard describes as an "off-the-peg fortune-telling kit."[75] The earliest version, in Greek, was probably created in the second century CE. Today, we know it through two papyrus versions from several centuries later. It included detailed instructions that make it easy to use, even today. Indeed, almost anyone with minimal theatrical skills and a bit of chutzpah could have used it to set up as a fortune-teller. Given both talents, you might even repeat, with conviction, the claim that the oracle was devised by Pythagoras, gifted to King Ptolemy by the sage Astrampsychus (a mythical magus, possibly from Persia), and used with resounding success by Alexander the Great. Impressive credentials! But we should take them seriously (if with a pinch of salt), because the oracle's longevity suggests that it met real needs successfully enough that many were willing to pay for its services. But the oracle also shows how divinatory methods could be used to dress up and give authority to forms of future thinking based on pragmatic experience and intuitions.

At the heart of the kit are ninety-two generic questions. Given the oracle's wide use, we can assume that the questions were winnowed over time to include the most common questions, those that could generate a decent income. The questions suggest that the oracle's main clients came from nonofficial sectors of the elite world. They were urbanized males, probably literate and moderately well-off, though some must have been (prosperous?) slaves because there are questions about the likelihood of being freed from servitude.[76]

To each question there are ten possible answers, which, like the questions, had surely been winnowed by a divinatory equivalent of natural selection over many years. They represented, in fact, a crude form of social statistics, pointing to the most common outcomes in familiar life situations.[77] Clients found the answer that applied to them by randomly picking a number between 1 and 10, and adding that number to the number of their question. The resulting total was used, after consulting a special key, to reach one of the ten possible answers to that question. Randomization allowed the gods to intervene because, as the oracle informs us, the number picked by the client is one "which god gives [the questioner] at the moment he opens his mouth." Such practices, or "divination by lot," were ubiquitous in ancient divination. Generally, you asked a question, threw a number of dice or knucklebones, and got an answer that depended on the total you threw.[78] Random dipping was a powerful tool in divination, not just because it opened a channel for divine intervention, but also because it could nudge your imagination in new directions.

Here's an example of how the oracle works. While writing this, I picked question 44: "Will I have a long life?" I then randomly chose the number 5. (It popped into my brain.) The two numbers add up to 49, and when I looked up 49 in the key, it sent me to the forty-fifth group of ten answers. There I looked up the answer corresponding to my random number 5 and found: "You won't have a long life. Put your affairs in order." Hmm!

The ninety-two questions tell us a lot about the anxieties that drove people to diviners. Most lurk within the Red Zone of chapter 2's Future Cone of Domains of Anxiety. This is where we find problems that worry us deeply (so we are willing to put effort, time, and money into resolving them), and where we suspect prediction may be possible (so it is worth seeking advice). There is no point in paying oracles to answer questions that do not worry us or are easy or impossible to predict. The Red Zone creates the itch that oracles scratch.

Questions fall into several groups. One concerns travel: Will

I travel safely? Will I ever leave this place? Some are more haunting: Will the traveler return? Is the traveler alive? A second group of questions is about careers: Will I serve in the army, perhaps, or become a general or a cleric or bishop, or a municipal official or perhaps a senator? Several questions concern business: Will this project prove profitable? Will I get my deposit back? Will I sell my cargo? Another group is about judicial proceedings: Am I safe from prosecution? Will I be released from detention? Will I defeat my opponent? Will I be caught as an adulterer? Some questions concern family, personal life, and health. Inheritances loom large because they were a major source of new wealth in the ancient world. Will I inherit from my father/mother/friend/wife, and will I get the dowry? This group includes the forlorn question: Will I have an inheritance from someone? Some questions concern marriage and family: Will I marry, and will it be to my advantage? Is my wife having a baby? Will she stay with me? One question — Will I rear the baby? — is probably for those thinking about exposing a baby at birth, which was a common way of avoiding scandal or even the cost of rearing unwanted babies. Some are about health, such as: Have I been poisoned?

Most of the oracle's answers hit the sweet spot between detail and generality that was described in chapter 2. They have enough detail to be interesting and plausible but are not so general as to be vacuous. Here are the possible answers to my own question about how long I will live: "You won't have a long life. Put your affairs in order" (repeated four times out of ten with minor variations); "You'll have an average lifespan. Don't be upset. Pray instead"; "You'll have a long life and suffer pains in your feet"; "After a time you'll succeed and grow old"; "You'll have a long life and wealth. You'll ask for more"; "You'll have a long life — and a very good one." Collectively, these answers cover most plausible futures. But they also contain just enough detail to be of interest, and even, perhaps, to risk being falsified.

Will I really suffer from pains in my feet? Sore feet were surely common among the old, but far from guaranteed.

The ancient historian Jerry Toner suggests that the oracle's answers were probably well enough tested that they offer a simple form of social statistics. About one-third of the answers to the question: "Will I rear the baby?" suggest that the baby will die or "not be reared" (a possible euphemism for exposure or abandonment). Modern scholarship suggests about one-third of babies died in the first year of life in the Roman Empire, so this looks about right. If Toner is correct, this suggests the intriguing probability that durable oracles may offer well-tested empirical evidence about social statistics in the ancient world. Aggregating the ten answers to question 12 — "Will I sail safely?" — suggests a 50 percent chance that sailing will be delayed and a 20 percent chance of serious danger, including shipwreck. The implied estimate of serious danger is similar to that found in other oracles, so it may be broadly realistic for the era.[79] These were the sorts of estimates that lay behind ancient actuarial calculations. And they serve as a reminder that future thinking that looks divinatory was often shaped by pragmatic realities based on awareness of real-world trends and everyday intuitions about the future.

All in all, there is much to recommend in the Oracle of Astrampsychus. Much of its predictive machinery is plausible and rational. The questions are serious. The answers are realistic, and they skillfully balance accuracy and detail. They may even have some statistical plausibility. Finally, as we have seen, randomization can be a perfectly rational tool for prediction. Despite its implausibility from a modern, scientific perspective, divination could offer consolation, and powerful and even reasonable forms of guidance.

CHAPTER 7

Modern Future Thinking

If man can, with almost complete assurance, predict phenomena when he knows their laws, and if, even when he does not, he can with high probability forecast the events of the future on the basis of his experience of the past, why, then, should it be regarded as a fantastic undertaking to sketch, with some pretence to truth, the future destiny of man on the basis of his history?

— MARQUIS DE CONDORCET, *SKETCH FOR A HISTORICAL PICTURE OF THE PROGRESS OF THE HUMAN MIND*, 1794[1]

The Modern Era of Human History

In the few centuries of the modern era — about one-thousandth of the time since humans evolved — changes have been even more spectacular than in the agrarian era. The emergence of global networks of exchange since 1500 lit the fuse for many of these changes. But the most spectacular transformations have occurred since 1800. Technological and scientific innovation soared as cheap energy from fossil fuels drove cascades of experimentation; global exchange networks brought more and more people into contact with new technologies and new ways of thinking about the world; and change occurred faster than ever before. In this brief period, human future thinking has changed more profoundly than in all earlier eras of human history.

These changes have taken us into a new phase of planetary history that many scholars call the Anthropocene epoch: the geological epoch in which, without planning to, humans started shaping the future of an entire planet.[2] That is why modern future thinking concerns itself increasingly with the future of all humans and all the other species with which we share the biosphere.

The following statistics hint at the astonishing scale of these changes. In just 220 years, between 1800 and 2020, the number of humans on Earth multiplied by almost nine times, rising from about nine hundred million to almost eight thousand million.[3] That is an average growth rate of about 1 percent per annum, or more than twenty times the rate of the agrarian era. Remarkably, most people are well-fed, thanks to an increase in the amount of land being irrigated and farmed, and to technologies such as genetic engineering and artificial fertilizers that have increased food production fast enough to keep up with soaring populations. Rising productivity in other areas made it possible (in principle) to house, clothe, and equip everyone to higher standards than ever before. The number of people living in cities rose from about 7 percent to almost 55 percent, meaning that cities have become our species' normal habitat. The populations of the largest cities increased from about one million to almost thirty million. Total human consumption of energy increased by about twenty-five times, from just over twenty thousand million gigajoules per year to roughly five hundred thousand million gigajoules. Energy use per person tripled, rising from almost twenty-five gigajoules per year to about seventy-five. Most of that energy came from a new source: the burning of fossil fuels, which is why carbon dioxide emissions increased by more than a thousand times in the same 220-year period, rising from just thirty million tons per year to more than thirty-six thousand million tons per year. There is one more remarkable statistic: the span of human lives increased. For most of human history, average life expectancy was below thirty years, though by 1800 more food and better healthcare had raised it to about thirty-five years. Between 1800 and 2020 the expected life span of each baby born on Earth doubled to seventy years.

New Technologies, Expanding Networks, and Accelerating Change

Modern technologies have given our species unprecedented power to shape and transform our futures, for better or worse. They allow us to communicate instantaneously over thousands of kilometers, to study objects smaller than a grain of dust or billions of light-years away, and to build machines that can carry us from one side of the world to the other in less than a day. And they have created new dangers. We are now burning fossil fuels on such a scale that we are transforming the atmosphere and oceans, and our weapons of war could ruin the biosphere in a few hours if we were foolish enough to use them. These changes in our ability to *manage* the future have been described many times in histories of technology.[4] So this chapter, like chapters 5 and 6, will focus on changes in *thinking* about the future.

The extraordinary scale of innovation has generated a new, collective hubris in modern future thinking. The modern idea of "progress" tells us we can remake the Earth for our own purposes, and the extraordinary wealth created by modern technologies makes that idea credible, because today, for the first time in human history, a majority of people no longer have to fight tooth and nail to survive. On the other hand, we have also become aware that our new powers have dangerous and unpredictable side effects that could lead to ruin. We have become so powerful that our future thinking is increasingly about how we humans will manage the fate of an entire planet and its many inhabitants.

Globalization and the creation of worldwide networks of exchange have also encouraged future thinking on global scales.[5] Before the sixteenth century, the largest human networks were those of Afro-Eurasia. Since then, traders and navigators, using methods both benign and brutal, have woven all human communities into a single planetwide network of almost eight billion people. We know of no other species that has ever been networked globally as humans are today. However, there are striking analogies here to the evolutionary processes that linked individual cells increasingly tightly within the first multicellular "macrobes."

Globalization destroyed as it created. From Siberia to Mesoamerica, from the Pacific to Africa, globalization brought European soldiers and Eurasian diseases that destroyed lives, societies, and economies and undermined ancient cultural certainties. In Europe, too, globalization shook up old knowledge, but here elites generally celebrated it because it also brought new forms of wealth, power, and knowledge. For a while, globalization turned what had long been a global backwater into the most dynamic, prosperous, and powerful region on Earth. European governments, traders, and scholars benefited from the creation of the first global networks because, for a few centuries, the region found itself at the center of global flows of power, wealth, and information. That explains why many of the changes we associate with modernity, including new technologies, new forms of economic management, and new forms of future thinking, emerged first in Europe and what came to be known as "the West," before being taken up and adapted in much of the rest of the world.

Globalization transformed understandings of time and the future as communities throughout the world found their lives were beginning to be shaped by global schedules that displaced traditional rhythms. Suddenly, Siberian reindeer herders and Pacific Islanders had to take note of the rhythms of warfare, trade, and taxation in distant imperial centers, while modern industry began to impose new rhythms on work and leisure, recreation and learning. Clocks became more precise, and some people began to carry private watches. In the eighteenth century, many watches had minute hands; in the nineteenth century, some had second hands.[6]

A single global clock time emerged. In the nineteenth century, governments and businesses began to synchronize clocks and calendars. In the 1840s, British railways published timetables synchronized to Greenwich Mean Time, and by the early twentieth century, most countries had aligned their time zones to Greenwich Mean Time. Calendars were realigned, too, as more countries adopted the Gregorian calendar. Traditional calendars, such as the Muslim lunar calendar or the traditional Chinese calendar, still shape the rhythms of billions of lives, but fireworks now mark the beginning of the Gregorian New Year in all major cities of the world. By 2020, almost everyone was caught in the mesh of a single global social time.

Finally, the pace of change has accelerated to such a degree that almost everyone now lives in a Heraclitean world of never-ending change. Nothing seems stable. Everyone faces the turbulence of A-series time. As the philosopher A. N. Whitehead argued in the 1920s, this change is momentous. "We are living in the first period of human history for which [the] assumption [of a fundamental stability] is false."[7]

Change has become so familiar that we can easily forget the strangeness of modern technologies. In 1829, at the age of twenty-one, the English actress Fanny Kemble was introduced to the revolutionary new technology of the railway by one of its founders, George Stephenson.[8] Living in a world of horse-drawn carriages, she instinctively thought of the engine as a mechanical horse, though she was stunned by its speed.

> This snorting little animal, which I felt rather inclined to pat, was . . . harnessed to our carriage, and, Mr Stephenson having taken me on the bench of the engine with him, we started at about ten miles an hour. . . . [Later] the engine having received its supply of water, . . . [it] set off at its utmost speed, THIRTY-FIVE MILES AN HOUR, SWIFTER THAN A BIRD FLIES (for they tried the experiment with a snipe).

I grew up before the era of space rockets, personal computers, the internet, and smartphones, but now I take these innovations for granted. The pace of technological change has dulled our sensitivity to the new.

We have also learned that change reaches deeper into the past than most of our ancestors could have imagined, and will continue further into the future. Before the modern era, most scholars assumed that, even if human societies seemed to change and evolve, the universe, the Earth, and the many species that lived on Earth had changed little since their creation. But from the seventeenth century, biologists and geologists, intrigued by the discovery of fossils and strange twisted or cross-cutting geological formations, began to understand that the Earth and the species that inhabit it (including humans) have changed profoundly on scales of hundreds of millions of years.[9] The heavens at least seemed changeless until the middle of the twentieth century, when astronomers found evidence that the universe, too, had a history. It was born in the fireball of the big bang and has been expanding and evolving for more than thirteen billion years. New chronometric dating techniques, first developed in the mid-twentieth century, even allowed us to construct astonishingly precise timelines for these histories, reaching back to the big bang.[10] The stable universe of ancient societies has been replaced by a turbulent, evolving universe in which we can be sure that the future will be different from the past.

New Understandings of Reality: Science and Disenchantment

Modern future thinking has also been shaped profoundly by the emergence of modern science.

We should not exaggerate the changes brought about by what has often been described as the "scientific revolution" of the seventeenth century. Many forms of divination are alive and well today, not least in the astrology sections of many newspapers and websites. My wife grew up with an old Balkan tradition that if two people walk around different sides of a lamppost or pole, someone must say "bread and butter" straightaway to avoid bad luck. Two people saying "bread and butter" is even better. And I should confess that I often touch wood to avoid bad outcomes, often as a joke, but secretly hoping it might make a difference. Nevertheless, the future-thinking methods embedded within modern science really are different, and they have transformed future thinking within many domains of modern life.

Modern science has several distinctive features. But, as historian Steven Shapin argues, the key shift was toward a more mechanical vision of how the world works, a vision that minimized the role of unpredictable spirits and forces in shaping the future.

A mechanical account of nature was explicitly contrasted with the anthropomorphism and animism of much traditional natural philosophy. . . . All seventeenth-century mechanical accounts set themselves in opposition to the tradition that ascribed to nature and its components the capacities of purpose, intention, or sentience.[11]

The founders of modern science, with Newton's laws of motion as a model, began to imagine a world governed by universal, mechanical, and impersonal "scientific laws" laid down by the one supreme deity. They expelled from their universe most of the spirits, demons, gods, and magical forces of the past, whose capriciousness had made prediction so difficult. In this more orderly, law-governed universe, it was hoped that new knowledge would offer powerful new ways of predicting and controlling the future.

The German sociologist Max Weber called this intellectual sea change the "disenchantment of the world," in a metaphor borrowed from the poet and philosopher Friedrich Schiller. (A more literal translation of Schiller's *Entzauberung* might be the awkward English word "demagification.")[12] At the heart of modern thought, according to Weber, is the idea of a "rational" world in which "there are no mysterious incalculable forces that come into play, but rather . . . one can, in principle, master all things by calculation. This means that the world is disenchanted. One need no longer have recourse to magical means."[13] The emergence of a more mechanical vision of the world had ancient roots in the increasingly impersonal divinatory techniques and the search for universal and impersonal laws of change that we saw in the religious and philosophical thought of the Axial Age.

Disenchantment did not lead straight to atheism, though many feared it would. Almost all the pioneers of modern science believed in a creator god, who had laid down the fundamental laws of the universe. Many, such as the philosopher and

scientist Robert Boyle, even accepted the existence of "an ines-
timable multitude of spiritual beings." But these thinkers
rejected the idea that spirits could arbitrarily mess with the fundamental laws of the universe. The astronomer Johannes
Kepler, for example, gave up the idea that planets had souls
and purposes and came to believe that "the Universe is not similar to a divine animated Being, but similar to a clock."[14] Clocks,
unlike many ancient gods, will not act whimsically or throw tantrums. You can predict their behavior, so you know what they
will do in the future.

The early successes of mechanical "natural philosophy" gave
it prestige and a brash confidence, and the new technology of
printing quickly spread this new way of thinking among European elites. As the historian David Wooton writes, in Shakespeare's time, even well-educated Europeans took magic and
witchcraft seriously. They believed in werewolves and unicorns.
They thought the heavens turned around the Earth; that comets were bad omens; that the *Odyssey* and *Aeneid* were true histories. A century and a half later, in Voltaire's time, educated
Europeans were fascinated by natural philosophy. Many used
telescopes or microscopes; they thought of Newton as the greatest of all scientists; and they knew that the Earth orbited the
sun. Though many remained superstitious, they did not take
magic or evil spirits too seriously, and they knew that unicorns
and miracles did not exist. Some even doubted the existence of
God, and many believed that the advancement of scientific
knowledge would lead to progress and a better future for
humanity.[15]

Today, though many still believe in a world full of spirits and
gods, the disenchanted worldview of modern science, spread by
mass education and the prestige of science's many successes,
shapes most forms of technological change and dominates
future thinking.

Future Thinking in a Mechanical Universe

Modern future thinking differs in four main ways from the future thinking of earlier eras:

1. *Causation:* A better understanding of causation allows more confident and precise prediction in many fields, including physics, chemistry, and medicine.

2. *Probability:* Probability theory offers a more precise understanding of processes in which we cannot predict specific events but can roughly predict the outcome of many events.

3. *Data Collection and Statistics:* A vast increase in the statistical information available, combined with new probabilistic methods, has enhanced our ability to detect, analyze, understand, and measure probabilistic trends that offer clues to likely futures.

4. *Information Technology and Computing:* Modern computer technologies allow us to store and analyze statistical information on previously unimaginable scales and with unprecedented speed and precision.

These changes have improved our ability to forecast in many domains, from medicine to demography and climate change. But in others, including domains such as politics that are shaped by the decidedly nonmechanical actions of humans, our future thinking is little better than that of the ancient world.

Causation

As we saw in chapter 2, understanding *why* things happen enhances our ability to detect and use trends in future thinking. Identifying trends is helpful, as Pavlov's dogs found when they learned that food arrives when the bells ring. But understanding the *cause* of trends is so much better! If we know *why* B follows A, we can predict the likely consequences of A with much more precision. Much of modern medicine is based on the nineteenth-century discovery, by scholars such as John Snow and Louis Pasteur, of *germ theory*: the idea that many diseases are caused by microorganisms. That meant that many diseases could be treated by keeping medical environments sterile, by vaccination, or with medicines such as antibiotics that attack the microorganisms. The historian Roy Porter comments: "In the century from Pasteur to penicillin one of the ancient dreams of medicine came true. Reliable knowledge was finally attained of what caused major sicknesses, on the basis of which both preventions and cures were developed."[16]

Understanding of causation is even more powerful if we can measure *how much* A affects B. This is why modern science loves careful measurement. Chewing willow leaves is an ancient remedy for headaches. It *caused* headaches to go away. Nineteenth-century chemists figured out why: the active ingredient in willow leaves is salicylic acid. Understanding that made it possible to manufacture pills containing salicylic acid, which were cheap and easy to use. What's more, you could measure the strength of each pill and figure out the difference between taking two and taking one hundred pills. Take two and your headache will probably go away. Take one hundred and you may die. These pills were known as "aspirin" in 1899, as they are today.

Many scientific advances depended on better understanding of causation. In 1644, the Italian mathematician Evangelista

Torricelli offered a mechanical explanation for the strange fact that a vertical tube filled with mercury and closed at the top would not drain completely if its open lower end was placed in a basin of mercury. Some mercury would always remain in the tube, leaving a vacuum at its upper end. Traditional (Aristotelian) explanations of this curious phenomenon depended on the idea that nature "hated" the vacuum and tried to keep it as small as possible. This was a purposeful, or teleological, explanation. Torricelli proposed a mechanical explanation. He suggested that mercury was forced up the tube by "air pressure," in other words, by the weight of the kilometers-high column of air pressing down on the open basin of mercury. In 1648, the French mathematician and philosopher Blaise Pascal tested Torricelli's idea by getting his brother-in-law to carry a similar apparatus up a mountain, the Puy de Dôme in central France. Higher up, Pascal reasoned, the weight of air should be less, which, if Torricelli was right, meant that the mercury in the tube should not rise as high at the top of the mountain. And that is what they found. Torricelli's apparatus was, in effect, a barometer: a way of measuring air pressure. The experiment converted Pascal to Torricelli's mechanical explanation of air pressure.[17] A measurable mechanical understanding of what caused air pressure would eventually allow the construction of steam engines, the foundational technology of the fossil fuels revolution.

Scientific explanations of causes depended on the assumption that many, perhaps most, processes are regular, mechanical, and measurable. Newton's laws of motion allowed you to predict the motions of cannonballs, planets, and falling apples with unprecedented precision. With that model in mind, modern science began to tease out new causal laws that allowed precise and measurable predictions in medicine, chemistry, and electricity, and eventually in areas such as nuclear physics. Increasingly precise understanding of causation lies behind most modern technologies, from smartphones to turbines, from jet planes to heart and lung machines. As the American statistician Nate Silver writes, "Predictions are potentially much stronger when backed up by a sound understanding of the root causes behind a phenomenon."[18] Judea Pearl, who pioneered modern perspectival ways of thinking about causation, writes that a good causal model allows you to know not merely "how things behaved yesterday, but also how things will behave under new hypothetical circumstances."[19]

If scientific understanding of causes allowed confident predictions in so many new domains, why not everywhere? Social theorists from Adam Smith to Auguste Comte and Karl Marx searched for causal laws in the evolution of human societies akin to Newton's laws of motion. But eventually, it became clear that not all domains of reality are shaped with equal precision by causal laws. That is one takeaway from the Future Cone of Domains of Predictability in chapter 2. Many domains, including the workings of human society, are shaped by looser causal laws. Asking what caused World War I is very different from asking what causes planets to move in elliptical orbits.

Probability

Modern probability theory arose from attempts to improve prediction in domains where causal links were looser and less mechanical than those Newton had found in astronomy and physics. "It is a truth very certain," wrote Descartes, "that when it is not in our power to determine what is true we ought to follow what is most probable."[20]

Hunches about probabilities, about the likelihood of dying in childbirth or returning from an ocean voyage, are ancient and universal. Modern probability theory uses mathematical models to place those hunches on a more precise footing. And, as poker players and insurance companies know, a more precise understanding of the odds can not only increase your under-standing of likely futures; it can make you a lot of money too. Probabilistic models are powerful because they often predict likely futures surprisingly well.

The roots of modern probability theory lie in the study of games of chance. Games of chance are ancient. Knucklebones that were probably used as dice have turned up in Bronze Age sites from the eastern Mediterranean.[21] But not until recent centuries were the probabilistic rules of games of chance studied in the mechanistic and mathematical spirit of modern science.

In 1564, an Italian mathematician, physician, and gambler called Girolamo Cardano wrote one of the first careful studies of games of chance, though it was not published until 1663. His book includes what Ian Stewart describes as "the first systematic treatment of probability." That the book was written was itself against the odds.[22] Cardano's mother tried to abort him. Though sickly, he survived, and even lived through a bout of bubonic plague that killed his nurse and brothers. In some ways his future thinking was not at all modern. When in trouble he admitted to visiting "diviners and wizards so that some solution might be found to my manifold troubles," and when gambling he often explained odd runs of losses by the fact that "fortune is averse."[23] But he was good at gambling, and, despite his superstitions, he thought through the logic of chance with mechanical precision, as if no imps or sprites or wizards were messing with the trends pointing to likely futures.

Here's an old problem that Cardano solved with mechanical precision. Experienced gamblers knew that if you threw three dice, they would add up to 10 slightly more often than they would add up to 9. That's the sort of difference gamblers thrive on — it is a counterintuitive idea, so it can often be used to win against beginners. But this is not a tight, causal law; it is probabilistic. In any particular throw, there is a good chance that you will get a 9 rather than a 10, but if you play long enough and keep betting on 10, you should do better than someone who keeps betting on 9.[24]

Why? Cardano offered an explanation that is easiest to understand by using the modern idea of a "sample space."[25] A sample space is a list of all possible outcomes of a process such as tossing a coin. But sample spaces can exist either in a model world, created inside our heads by the firing of billions of neurons, or in the real world. And that distinction is crucial to all probability thinking. Sample spaces in the world of models can be known completely and their behavior can be described with mathematical precision. Those in the real world are less well-behaved. If you toss a coin in the model world of your imagination, the sample space is simple: it consists of one head and one tail, with each outcome having a 50 percent chance. In the real world, your coin could be old and battered so there might be a slightly higher chance of it turning up heads.[26] Real-world sample spaces are messier, and we never really know exactly what they contain. But if real-world samples look pretty much like our models, we cross our fingers and hope the models can give us pretty good guidance about the real world. Surprisingly often, that bet pays off.

To solve the problem of 10s and 9s, Cardano constructed a model sample space for all the possible totals you might get if you throw three dice. In the real world, you can only throw once, but in the world of models, you can play the same game many times and see all possible outcomes. With three dice, there are $6 \times 6 \times 6$ or 216 different possible outcomes, and we assume that, with fair dice (we're in the model world, not the real world so far), each one has the same likelihood of turning up. Among those 216 different throws you'll find six different ways of getting a total of 9, and six different ways of getting a 10. For example, you can get a 9 by throwing 6,2,1, or 5,3,1, or 5,2,2 or 4,4,1, or 4,3,2 or 3,3,3. Shouldn't that mean that both totals are equally likely? No. Look at the list more carefully and you'll find that not all six ways of throwing a 9 are equally likely. There's only one way of getting a 9 by throwing three 3s, but you can get a 9 in six different ways if you throw a 6 a 2 and a 1, but in different orders (6,2,1 or 6,1,2 or . . .).[27] Count the possibilities, as Cardano did, and it turns out there are twenty-seven ways of rolling a total of 10 (giving you a 27/216, or 12.5 percent, chance) but only twenty-five ways of rolling a total of 9 (giving you a 25/216, or an 11.6 percent, chance). There's the crucial difference! And, with this explanation in mind, you can make some money. Assuming, of course, that the real-world dice are not loaded, and that most of the time the real world is pretty similar to the model world.

Cardano's ideas had little impact outside the world of gambling. But in the middle of the seventeenth century, Blaise Pascal and others floated the powerful new idea that careful, Cardano-type thinking about probabilities might help not just with games of chance, but in many other realms of future thinking. Could we construct a model sample space based on past experience that could help trading companies calculate the probability of a ship foundering? Could we even solve metaphysical problems such as the existence or nonexistence of God? These exciting ideas emerged in the middle of the seventeenth century and were quickly taken up by leading European thinkers.

In 1654, an aristocratic gambler known as the Chevalier de Méré put to Pascal and fellow mathematician Pierre de Fermat a problem about how to divide the stakes in a game of chance if it were interrupted when each player had already attained a certain number of "points." This is the "problem of points." Cardano had also tackled it, but Pascal's and Fermat's solutions took probabilistic calculations to a new level of sophistication.

Pascal's solution depended on "a full enumeration of all possible outcomes" in an imagined sample space.[28] The mathematics are beautiful and elegant, but they miss the messiness of the real world, so Pascal's work illustrates the dangers of probabilistic thinking. For example, Pascal had to assume that each player would continue to play at exactly the same level of skill as in the games already played, with no effects from drink or tiredness or tension. Removing real-world contingencies left Pascal with a perfect mechanical model, from which all caprice had been expelled. What's more, in the model world, card games or horse races or wars or climate changes can be replayed many times to find out which outcome is most common and therefore most likely. But reality is rarely this neat and real card games are played just once. As Warren Weaver puts it, "What one does in probability theory, is invent a mathematical model, which can be calculated in a completely clear and tidy way, and then hope that this model will correspond in a useful way to some real phenomena."[29] Probability theory cannot avoid the inelegant inductive leap of faith present in all future thinking (and discussed in chapter 2). But it can make the logic behind that leap more transparent, and even measurable, and that can add precision to future thinking whenever we are pretty confident that our models really do capture important aspects of the real world.

Pascal's famous "wager on the existence of God" illustrates even better the dangers of unrealistic models. In 1654, Pascal underwent a profound religious crisis, and in subsequent notebooks (published as the *Pensées,* or *Thoughts*) he extended probabilistic calculations to theological and metaphysical problems. His wager on the existence of God turned a theological question into a sort of bet. Pascal set up a model sample space with just two possibilities. Either there is no god, or there is a Christian God who promises eternal salvation for the good and eternal damnation for the bad.[30] Then Pascal makes another questionable assumption: he guesses that each possibility has a 50 percent chance of being correct. "A game is being played . . . where heads or tails will turn up." Because the chances are equal, we need to look at the payoff to the two outcomes before placing our bets. Act as if God does *not* exist, and you may have fun for one brief lifetime, but you risk an eternity of misery if you lose. Act as if God *does* exist, and at worst you will miss out on the fun of the sybarite for one lifetime, but the possible gain is infinite happiness in the afterlife. "Let us weigh the gain and the loss in wagering that God is. . . . If you gain, you gain all; if you lose you lose nothing. Wager, then, without hesitation, that He is." The logic is impeccable and striking, but how plausible is Pascal's imagined sample space? Not very!

In 1662, colleagues of Pascal offered a more realistic defense of probability theory in the final chapters of *Logic, or the Art of Thinking*, a new text on logic that largely replaced Aristotle's *Logic* in European universities until the nineteenth century. They argued that probabilistic logic could clarify reasoning about likely outcomes in the real world.

> Many people [they wrote] . . . are exceedingly frightened when they hear thunder. . . . But if it is only the danger of dying by lightning that causes them this unusual apprehension, it is easy to show that this is unreasonable. For out of two million people, at most there is one who dies this way. . . . So, then, our fear of some harm ought to be proportional not only to the magnitude of the harm, but also to the probability of the event. Just as there is hardly any kind of death more rare than being struck by lightning, there is also hardly any that ought to cause us less fear, especially given that this fear is no help in avoiding it.[31]

Having grown up with a great-aunt who locked herself in the toilet as soon as she heard thunder, I understand these fears. But I also appreciate the clarity that careful probabilistic thinking can bring to them. The sample space of causes of death proposed in this passage is based on past experience, and the assumption that in the future, too, only one person in two million will die from lightning is both plausible and enlightening. In other words, past trends show that death by lightning is far too unlikely to fall within the Red Zone of the Future Cone of Domains of Anxiety. Probability theory is, as Laplace put it two centuries later, a "calculus" of probabilities, but we must always remember that we make a leap of faith in applying that calculus to the real world.

Over the next three centuries, the mathematics behind probability thinking would become increasingly refined. In *The Art of Conjecture,* first published in 1713, eight years after the author's death, Jacob Bernoulli showed that you can reverse the logic of probability thinking. Instead of asking what the odds of a given outcome are, you can look at several outcomes and ask: What can these outcomes tell me about the sample space they are drawn from? This is "inverse probability," and it is a powerful way of gaining rich insights from limited information, including insights about likely futures. Inverse probability makes inferences about whole populations on the basis of samples. Political pollsters use it all the time as they try to predict election results from a few interviews.

Bernoulli imagined a model world with an urn containing hundreds of black and white tokens. If we pick ten at random, and six turn out to be white, what does this tell us about the proportion of white and black tokens in the whole urn? In other words, what proportions will I get if I keep picking tokens from the urn? Can I assume that more or less 60 percent of the tokens will turn out to be white? Bernoulli proved mathematically that the larger your sample, the better it approximates to the underlying distribution. This is the "law of large numbers" and it makes intuitive sense because eventually your sample could include every token in the urn, at which point your sample will be the same as the underlying distribution. Less intuitive is his finding that you can get good approximations of the underlying proportion of white and black tokens well before you have counted every token in the urn. Indeed, the closeness of your sample to the underlying population depends not on the size of the underlying population (which would require vast samples for very large populations) but on the size of the sample itself. That is a remarkable result, and the underlying justification for most forms of statistics that draw conclusions about large populations from limited samples.[32]

Inverse probability teases out the mathematical logic behind the ancient idea of random dipping. Random samples give us limited knowledge of the world, but pollsters know that samples of a few hundred or a few thousand can usually give us predictions that are quite accurate enough in practice. Of course, the samples must be picked as randomly as possible if they are to behave like the mathematical models. (Don't select your political interviewees from a group all wearing rosettes for the same political party!) Today, the idea of inverse probability is used most extensively in Bayesian statistics. As we saw in chapter 3, this approach begins with an initial, often highly subjective, estimate of the shape of a possible sample space, and updates that estimate as new information comes in.[33]

In the eighteenth century, other mathematicians, including Laplace, showed that you can even give mathematical estimates of how close your sample answers are to the real distribution. You can construct mathematical models of how random samples vary, and these allow you to estimate how close a sample probably is to the underlying distribution. For example, many real-world samples seem to vary according to a pattern often known as the normal distribution. It is sometimes called the bell curve because it is shaped like a bell. You see normal distributions if you list variations in the number of heads and tails that turn up in many different games, or the height of potential army recruits, or the number of days of extreme heat or cold. In a normal distribution, most outcomes cluster around the mean, or average. Then the number of outcomes falls away the farther you get from the mean, and they do so in ways that can be modeled mathematically. In normal distributions, the average variation of sample means from the population mean is measured by the standard deviation. In the world of models, 68.2 percent of all sample means from a normal distribution of all possible sample means will lie within one standard deviation of the population mean, and 95.4 percent will lie within two standard deviations. That knowledge allows you to say that there is a 68.7 percent chance that the mean for your sample lies within one standard deviation of the mean for the whole population.

How well does the real world match these neat model distributions? The answer is: well enough to make such models very helpful. The chart below gives height measurements for 36,658 eighteen-year-old recruits to the British army in the years 1880–1884.[34] The distribution is skewed by the fact that recruits shorter than 65 inches were normally rejected, though clearly

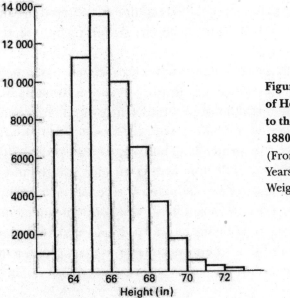

Figure 7.1: Distribution of Heights of Recruits to the British Army, 1880–1884
(From Rosenbaum, "100 Years of Heights and Weights," 281)

some slipped through the net. Without that bias, the distribution would look even more like a standard normal curve. The mean height for this group was 64.7 inches, and the standard deviation was 2.34 inches, which allows us to say that just over 68 percent of recruits probably fell within 2.34 inches of the mean and just over 95 percent fell within 4.68 inches. The chart shows the spooky way in which real-world distributions seem to mimic the world of models. And that explains why it is so tempting to project normal distributions into the future, for example, to predict the likely range of heights of recruits in a few years' time.

In the nineteenth and twentieth centuries, the mathematics of probability became increasingly sophisticated. But the most fundamental change was in how probability theory was interpreted. The "classical" probability theorists of the seventeenth to the nineteenth centuries assumed that the world worked deterministically, according to scientific laws, most of which were yet to be discovered. So they saw probability theory as a way of coping with ignorance. Before the nineteenth century, few scientists took seriously the idea that there might be genuinely random events. As Hume put it, "'Tis commonly allowed by philosophers that what the vulgar call chance is nothing but a secret and conceal'd cause."[35]

However, as we saw in chapter 1, modern science has largely abandoned Laplace-style determinism. It accepts that many events, such as the breakdown of a radioactive atom, are genuinely random. They do not have a "conceal'd cause," so they are unpredictable even in principle. That means that probability theory is not just a way of dealing with ignorance; it is the most accurate way we know to describe many aspects of reality. General laws may shape the universe at large scales, but the moment-to-moment coming into being of the universe and our own futures is probabilistic. Einstein was one of the last prominent defenders of a deterministic universe. According to a famous anecdote, he once told the physicist Niels Bohr that God does not play dice with the universe. Bohr supposedly replied, "Einstein, stop telling God what to do!"[36]

Data Collection and Statistics

In many situations, probability theory can increase the precision with which we predict likely futures. But only if we have plenty of information from the real world, such as lists of the heights of tens of thousands of recruits. And the more information the better. Bernoulli's law of large numbers explains why it is worth collecting a lot of information. More information means more detailed and precise knowledge of the evolution and shape of the long-term trends that lie at the heart of good future thinking. And it allows the use of fancier probabilistic models. So, the third distinctive feature of modern future thinking is the gathering of vast amounts of information within the modern discipline of statistics. Today, statistical thinking is ubiquitous and shapes crucial decisions made by governments, businesses, and scientists as they decide what new infrastructure or start-ups or research projects to invest in.

The roots of modern statistics go back to the seventeenth century. In a book published in London in 1662, a pioneering demographer John Graunt drew up the first life tables on the basis of weekly lists of christenings and burials over the previous sixty years. From this rich collection of information, Graunt and his colleague, the economist William Petty, teased out some tantalizing and important probabilistic conclusions. They offered estimates of the real population of London, gender ratios, the numbers of people who died at different ages, the number of possible recruits for the army, migration to and from London, and the impact of different diseases.[37] Such estimates were new, and of great interest to governments, which were keen to foresee and perhaps even control the future at large scales.

In the eighteenth century, there was a craze for statistics, driven in part by the discovery of unexpected regularities in human societies. The philosopher Ian Hacking suggests that the first law of modern social statistics was John Arbuthnot's discovery, in 1710, that about thirteen boys were born for every twelve girls.[38] That was unexpected new knowledge based on statistical data. If you are expecting a child, the odds of having a boy or a girl are *not* even. Was it possible that, beneath the apparent chaos of human social and biological behavior, there lurked many such patterns that might be used to forecast what had once seemed unforecastable? Could many aspects of human behavior be moved, perhaps, from the possible into the plausible or even the probable domains of prediction? Perhaps, but only after the collection of lots of information. In 1796, Mme. de Staël wrote:

> It has been discovered in the Canton of Berne, that the number of divorces is very much the same from one decade to the next, and there are cities in Italy where one can calculate exactly how many murders will be committed from year to year. Thus, events which depend on a multitude of diverse combinations have a periodic recurrence, a fixed proportion, when the observations result from a large number of chances.[39]

The hope that probabilistic mathematics applied to large amounts of social information might improve predictions about the future of human societies was immensely appealing to governments, entrepreneurs, economists, and social theorists.

In the early nineteenth century there began what Ian Hacking describes as an "avalanche of numbers."[40] Governments, officials, and scholars began to collect huge amounts of data as they searched for patterns in population growth, in criminality, in the spread of infectious diseases, in the weather, in economic change, and, in general, in the evolution of large, complex systems. Of course, ancient empires had all collected harvest data and held censuses. What was new was the sheer amount of information collected and published, the range of questions asked, and the mathematical sophistication with which that information was analyzed. More and more "laws" were discovered. Adolphe Quetelet, one of the pioneers of modern statistical thinking, was astonished by their regularity. "We know in advance," he wrote, "how many individuals will dirty their hands with the blood of others, how many will be forgers, how many poisoners, nearly as well as one can enumerate in advance the births and deaths that must take place."[41] Was it possible that human society might be subject to laws as regular as those of astronomy?

In the twentieth century, the limitations of grand theories of social development such as those of Comte or Marx became apparent. But at more modest scales, social statistics can provide guidance on matters of great significance, for example, in patterns of criminality or the incidence of different diseases or the likely need for different types of infrastructure. So the collection of social, economic, medical, criminological, and other types of social statistics has gathered pace to the present day. Statistical information is used throughout the world to guide investments, preparations for pandemics, management of economies, and the understanding of complex systems such as the global climate regime.

The IT revolution of the late twentieth century and the creation of the internet increased the amount of data that could be collected, stored, and analyzed. It became possible to tease out important trends not just from samples, but from complete sets of data. In the twenty-first century, the era of "big data," information about the tastes of individual consumers, their spending patterns, and their activity on social media (every "like" is noted and recorded) has been used to make huge profits because it predicts consumer demand with astonishing precision.

Though often associated with marketing, the phrase *big data* was coined in sciences such as astronomy and genomics, where it first became clear how huge collections of data could transform what was knowable and predictable. The Sloan Digital Sky Survey, which began in 2000, collected more astronomical data in a few weeks than all the information accumulated previously in the entire history of astronomy.[42] The power of big data arises from the synergy between vast amounts of information and new analytical techniques that can use that information to tease out hidden patterns and trends: "It's about applying math to huge quantities of data in order to infer probabilities: the likelihood that an email message is spam; that the typed letters 'teh' are supposed to be 'the.'"[43] Like whispers from the gods to ancient diviners, hints about hidden trends buried in mountains of data can now be turned into plausible predictions . . . and into wealth and power.

Information Technology and Computing

The ability to store, access, and mine huge amounts of data was made possible by the information revolution of the late twentieth century.

Modern electronic computing began in earnest during the Second World War. But for some time, computers looked like expensive and lonely machines, of value only for governments, the military, and large corporations. In one of the modern era's great failed predictions, the president of IBM said in 1943, "I think there is a world market for maybe five computers."[44]

Cheap and ubiquitous information storage and processing was made possible by technological changes that cheapened computer chips and let computers talk to one another. In 1965, Gordon Moore, cofounder of the chip maker Intel, noted that the number of components on each chip seemed to be doubling every year, and that was driving down the cost of computers and information processing. Though he scaled his prediction back to once every two years in 1975, the trend has held since then and is known as Moore's law. In 2020, a typical smartphone had a thousand times more computational power and eighty times as much storage as the best 1970 computer, but at one-thousandth of the cost.[45]

As computers became cheaper, more powerful, and more connected, it became possible to store and analyze colossal amounts of information, and to model possible futures with unprecedented rigor. A pioneering work on likely global futures, *Limits to Growth,* was published in 1972 by the environmental scientist Donella Meadows and her colleagues at the Massachusetts Institute of Technology. *Limits to Growth* made use of one of the first computer models of the entire planetary system and showed the extraordinary contribution such technologies could make to global future thinking. As the futurist Wendell Bell notes, no earlier study "had dealt simultaneously with so many key variables relevant to the survival of humankind into the far future, their interrelationships, and their consequences holistically, comprehensively yet simply, and so persuasively."[46]

The computer model used in *Limits to Growth* was designed by Jay Forrester at MIT and known as World3. It modeled links between the land, the oceans, the atmosphere, the natural environment, and human societies as dependent parts of a complex global system. It concentrated on five trends: population, industrial output per capita, per capita production of food, reserves of nonrenewable resources, and the impact of pollution, including greenhouse gases. These trends and their many subtrends were connected by multiple causal links and feedback loops, each of which could, in principle, be quantified from existing data or estimated plausibly. The original model included more than one hundred causal links. The authors ran their models many times, with slight alterations to key parameters such as rates of population growth or the availability of nonrenewable resources. They tested the model's realism by starting their runs in 1900 and watching how well the model tracked changes up to the publication date (1972), before seeing what it predicted up to the year 2100. In the original publication, almost all runs, particularly the "Business as Usual" scenarios, led to an eventual slowdown and collapse of growth in the twenty-first century, though the timing, the scale, and the proximate causes of the decline varied. Twenty years later, in *Beyond the Limits,* the authors found versions that avoided collapse, but only by incorporating large-scale, determined, and systematic decisions by human societies to limit production, population growth, and pollution.[47] The scenarios described in these two books, though simple by today's standards, have stood the test of time well, and most point to a flattening, and perhaps even a decline of many forms of growth by the year 2100.[48]

As computer technology evolved and spread, such modeling became more common and more powerful. Today, computer models with a lot more information and computational oomph are used to explore possible futures on many different scales, from climate change to the evolution of economic trends and pandemics.

The Power and Limits of Modern Future Thinking: Weather and the Economy

Modern methods of future thinking have increased our ability to predict in many domains of reality, particularly those shaped by fairly regular mechanical or probabilistic processes. In some areas, such as the prediction of asteroid impacts, they have shifted problems from the "possible" to the "plausible" and sometimes even the "probable" domains of chapter 2's Future Cone of Domains of Predictability. A 1994 order from the US Congress to NASA launched the systematic study of the likely trajectories of all near-Earth objects larger than one kilometer in diameter.[49] By 2020, about 95 percent of asteroids this size or larger had been tracked, and none seem likely to hit the Earth in the next hundred years. That is a real gain in predictive knowledge. There have been similar gains in many fields, including influential predictions about the public health impacts of behaviors such as getting vaccinated or smoking or wearing seat belts in cars. But in domains shaped by less predict-able processes, the future remains as obscure as in the past. Forecasting the actions of politicians has not advanced much since Caesar's time.

Modern weather forecasting and economic forecasting illus-trate some of the strengths and weaknesses of modern future thinking.

Weather forecasts use all the advances we have noted. They are based on better understanding of causation — of how changes in air pressure, moisture, and temperature drive changes in the weather. They use rigorous probabilistic calcula-tions to estimate the likelihood of different weather patterns. They use vast amounts of information from around the world. And they model likely changes in the weather using the immense storage and processing power of supercomputers.

All societies have tried to forecast the weather, and short-term forecasting is not hard. If there isn't a cloud in the sky, I'll confidently predict that it won't rain in the next five minutes. But predicting the weather tomorrow or next month is much trickier. Modern weather forecasting really began in the nine-teenth century. In the middle of the century, Captain Robert FitzRoy, commander of the HMS *Beagle,* on which Darwin voy-aged around the world, began collecting weather data from many different stations, and in 1854, he founded the English Meteorological Office. In 1875 the London *Times* published the first map of weather across the British Isles based on information from many local stations.[50] The American scholar Cleveland Abbe first suggested that weather systems could be modeled on the dynamics of fluids, and a Norwegian, Vilhelm Bjerknes, suggested that differences in local air pressure drove the currents of these atmospheric fluids. That meant it should be possible to forecast weather patterns by collecting and analyzing many local measurements of air pressure, temperature, wind speed, and humidity.[51] Modern forecasters also collect information from planes and satellites. In 1950, the mathematician John von Neumann developed the first computer-based weather forecasts, and soon after, Edward Lorenz began to build programs to simulate global weather patterns.

Today, meteorological information is collected from more than four thousand stations linked within a Global Observing System, as well as from orbiting satellites. That information is collected and processed in centers such as the European Centre for Medium-Range Weather Forecasts (ECMWF) in Reading, England. Forecasts using all this information are generated using elaborate models and the colossal computational power of supercomputers. In the idealized world of models, tomorrow's weather can be rerun thousands of times with tiny variations, and that makes it possible to say, for example, that there is a 50 percent likelihood of rain tomorrow. To test the reliability of such predictions, they are constantly checked against real outcomes so that weather forecasting is subjected to never-ending tests of its reliability. In 1979, when the Reading center was established, its two-day forecasts were worth taking seriously. By 2015, the center was offering equally reliable six-day forecasts. By 2025 it hopes to offer trustworthy two-week forecasts.[52] Though modest in their claims and probabilistic in form, such forecasts can be immensely important, not just for planning family picnics, but also in giving early warning of catastrophic weather events such as hurricanes or floods. Global warming will make such warnings increasingly important.

Modern forecasting methods have also been applied to economic change. Like weather forecasting, economic forecasting is extraordinarily complex because it is full of chaotic processes. In addition, though, it is shaped by the highly unpredictable actions of human beings. Much economic theory has claimed that human economic behavior is predictable in the aggregate. But it is also clear that a lot of economic behavior is *not* that predictable, including the behavior of the governments that set many fundamental economic parameters. Further, economic forecasters, unlike weather forecasters, are part of the system they are trying to predict, particularly if they are paid by governments and corporations that are actively trying to shape economic futures. That adds feedback loops that muddy attempts at prediction. Like ancient soothsayers, economic forecasters often know what forecasts their clients want.

The wars of the early twentieth century forced all governments to attempt economic planning, and that made governments and businesses increasingly dependent on economic forecasts. The failures of the Soviet attempt at total planning showed some of the limits of economic forecasting, but all modern governments try to steer their national economies. Not surprisingly, economic forecasting, and economic theory in general, have been shaped by political pressures and ideologically biased models of economic change. So much rides on economic forecasting that there is a premium on overconfident forecasting, which explains why so many economic forecasts are overprecise, so they miss the sweet spot between generality and precision. "We predict growth of 0.5 percent in the next three months" is very different from "40 percent likelihood of rain within the next twelve hours." Taken together, political bias, chaotic processes, and the unpredictability of human behavior help explain why so few economists foresaw economic earthquakes such as the 2008 global financial crisis.

In a scathing survey of failures in economic forecasting in the United States, Nate Silver takes as an example annual forecasts of US GDP growth in the coming year from the Survey of Professional Forecasters for 1968 to 2010. They give 90 percent ranges, which means that the actual outcomes should fall within the predicted range 90 percent of the time. But when checked after the fact, the predictions fell outside that range almost half the time, despite the fact that the 90 percent bands were already so wide as to be of little value. Predict 2.5 percent growth next year with a 90 percent error band and what you may really be saying is: "We predict something between a 5.7 percent growth and a 0.7 percent decline in GDP."[53] Not that helpful.

Future Thinking in the Modern World

How different is modern future thinking? Two thousand years ago, in his Platonic dialogue on *Divination*, Cicero wrote one of the most illuminating accounts of future thinking in the agrarian era. If we could resuscitate Cicero (unlikely because his head was cut off in the proscriptions that followed Caesar's assassination), what would he make of future thinking in today's world?

He would find much that is exotic and new. But many aspects of modern future thinking would seem surprisingly familiar. He would be shocked by the expulsion of the gods from official future thinking, despite his skepticism about popular religion. But he would note with approval that most modern governments still support and respect religious practices along with the worship of different gods. He would note with less approval that divination and astrology are still ubiquitous in popular future thinking. Given his rational and empirical turn of mind, he would be fascinated by the empirical and mechanical methods of future thinking described in this chapter, though he might struggle to understand some of them and the technologies behind them. He would note the immense importance of particular future-thinking skills in politics, business, science, and many other areas of modern life. And he would be impressed by the remarkable successes of future thinking in fields such as medicine and science. But he would also note (perhaps with some glee) that the track record of future thinking in politics is dismal, hardly any better than that of Roman diviners and augurs.

Finally, Cicero might be surprised to find that, while particular future-thinking skills are widely respected, such as those of the statisticians, computer modelers, scientists, and economic planners, who are today's soothsayers and seers, future thinking as a general domain of knowledge remains as fragmented and disreputable as in his own era. Where are the Nobel Prizes for future thinking? Or the professional organizations that could give formal accreditation to future thinkers? Or the school syllabi designed to help every student think more skillfully about the future? Particular future-thinking skills may be taught and admired, but future thinking in general remains as far from mainstream thought as in Cicero's time.

This odd mix of respect and neglect is evident in the history of the modern scholarly field of futures studies.[54] Future thinking loomed large within the planning agencies and scholarly institutions of the Cold War era. On both sides of the Iron Curtain, governments, planners, and scientists held high hopes for future studies because they shared a commitment to "progress" and believed that science would make rigorous predictions possible in more and more domains of reality. In the English-speaking world, H. G. Wells, one of the pioneers of modern science fiction, was also one of the first advocates for future studies.[55] In 1902, he argued confidently that "the events of the year A.D. 4000 are as fixed, settled, and unchangeable as the events of the year 1600." Advances in modern science meant that "a working knowledge of things in the future is a possible and practical thing," so that it should soon be possible to attempt "a systematic exploration of the future."[56] Soviet governments showed a similar optimism in their attempts to draw up rational plans for the future of an entire society. And capitalist governments also began to plan systematically for better futures.

The wars of the early twentieth century gave a military edge to much official future thinking as governments prepared for future wars. Nuclear weapons and the missiles that carried them were a product of militarized future thinking. In the mid-1960s, the advent of computers encouraged a new wave of optimism about the possibilities of rigorous, scientific future modeling. In 1964, members of the American RAND Corporation (RAND is a contraction of *research and development*) argued that it should soon be possible to deal with socioeconomic and political problems "as confidently as we do with problems in physics and chemistry."[57]

Optimism about the future of future studies peaked in the 1970s and 1980s. By the mid-1970s hundreds of futures courses were being taught in North America, and by the 1980s, hundreds of books on the future were being published each year.[58] The business world, too, fell in love with futures studies. Management consultant Peter Drucker led the charge for systematic business planning for the future, and large companies hired "futurists" to do their predicting.[59] Some scholars began to argue for a new "general theory of the future," and a German American scholar, Ossip K. Flechtheim, coined the term *futurology* in the 1940s.[60]

In the 1980s, these hopes began to ebb. As we saw in chapter 1, confidence in the predictive power of science waned as twentieth-century science became more probabilistic and less deterministic, while the collapse of the Soviet Union revealed the limits of economic and technological planning for whole societies. The emerging field of futures studies was also tainted by its close association with military and political goals. New forms of futures studies emerged at a distance from governments and government-funded think tanks, and they put less emphasis on prediction than on the exploration of popular hopes and fears for the future. In 1953, Fred Polak's *The Image of the Future* argued that futures studies should focus not on the future but rather on contemporary *images* of possible futures. In 1960, Bertrand and Hélène de Jouvenel created an Organization Futuribles, committed to the idea that futures studies should not predict, but should instead clarify the choices we make between alternative possible futures.[61] In 1968, the Turkish American scholar Hasan Özbekhan and the Italian industrialist Aurelio Peccei created the Club of Rome to support careful thinking about the future of humanity as a whole. *The Limits to Growth* was its first report.[62] In 1973, with UNESCO support, the World Futures Studies Federation (WFSF) was established to create a field of futures studies that could represent the future goals of all humans. That meant exploring multiple futures, which is why, today, many futurists prefer to label their field *futures studies*.[63]

In practice, of course, a field known as *futures studies* could never avoid prediction. As Wendell Bell argued, even those who resisted the word *prediction* continued to use "different euphemisms for the same idea, such as 'foresight,' 'forecast,' or 'projection,' to describe their work."[64] These quibbles show how the modern field of futures studies has had to search for the same sweet spot that all future thinkers have sought between overly specific and overly general visions of possible futures.

Futures studies have not become mainstream in modern schools and universities. Though particular types of future thinking are ubiquitous, the scholarly field of futures studies is subject to the same sort of intellectual skepticism that Cicero directed at diviners in his own time. Why? Much scholarship in the field, such as Wendell Bell's classic two-volume study of the field, is superb, creative, and rigorous. And there is a whole world of futurists and forecasters who work mainly for businesses, corporations, and government institutions, using a wide range of well-honed forecasting techniques such as scenario planning, backcasting, and the Delphi technique, which builds consensus positions among expert forecasters.[65] The skepticism may reflect the curious half-existence of the field's subject matter — the future — a topic on which we have no hard evidence and whose very existence can be doubted. Without documents from the future, it is hard to know what *rigorous, evidence-based* study of the future really means. We never really know in a given situation what alternative futures are possible, so we can never know if a successful prediction was accurate or lucky. No wonder that a whiff of metaphysics, mystery, and even skullduggery still surrounds future thinking and ensures a lingering skepticism about the field. The lack of hard evidence also explains why the borders between future thinking and fiction are so porous. Indeed, it is striking that some of the most interesting modern future thinking can be found not in futures studies, but in science fiction.[66] Because the clues we have to likely futures are so few and so slippery, creativity and imagination play a much greater role in future thinking than in historical research, which is tightly constrained by detailed evidence. That, too, invites skepticism about future thinking as a field of scholarly research.

And yet . . . despite the difficulty of thinking rigorously about possible futures, we have to try. We have no choice. And our efforts really matter. Nothing suggests more powerfully why careful and creative future thinking is so urgent than current debates about the future of humanity. These are the subject of the next chapter.

PART IV

Imagining Futures

Human, Astronomical, and Cosmological

CHAPTER 8

Near Futures

The Next One Hundred Years

Our Earth is forty-five million centuries old. But this century is the first in which one species — ours — can determine the biosphere's fate.

— MARTIN REES, PRESIDENT OF THE
ROYAL SOCIETY AND ASTRONOMER ROYAL, 2018[1]

What wouldn't we give sometimes for Krishna's god's-eye view of the future, with its breathtaking four-dimensional maps of B-series time! But in today's more disenchanted world, we have mostly given up on divination. We rely instead on the messy probabilistic styles of trend hunting described in chapter 7. These depend on better understanding of causation and probability, and the collection and processing of huge amounts of statistical information to identify past trends that hint at likely futures. Our maps of possible futures lack the serenity, precision, and certainty of Krishna's vision. They are tentative, blurry, speculative, lacking in detail, and may sometimes be as fantastical as medieval maps of unknown lands. But they are the only maps of the future we have, and that alone lends them importance and even a certain grandeur. All of them aim at the forecaster's sweet spot between excessive precision (which almost guarantees they will be wrong) and excessive generality (which makes them more or less useless).

Chapter 8 tries to imagine "near futures" over the next one hundred years. Chapter 9 peers into "middle futures," focusing on future stories for our own lineage, over thousands, even millions of years. In chapter 10, we try to imagine plausible futures for the Earth, the solar system, the galaxy, and . . . the universe as a whole.

Distinctive Features
of the Hundred-Year Time Scale

Each of these future time scales has distinctive features. There are trends in the next hundred years that can be predicted with some confidence, because the near future is close, and we can see some of the more regular trends that will shape it. Nevertheless, even the near future lies mostly in darkness. There are so many unknowns, and so many critical choices will be made by members of that most unpredictable of species: human beings. The hundred-year future is personal, because it will be inhabited by people we know and care about, so it falls under what Elinor Ostrom calls the "seven generation rule." This is the idea, familiar to many indigenous peoples, that "when we make really major decisions, we should ask not only what will it do for me today, but what will it do for my children, my children's children, and their children's children into the future."[2] We humans have become so powerful that, whether we understand what we are doing or not, our activities will shape the near future. How we imagine the future today will shape the decisions we make tomorrow, and those may haunt planet

Earth for millions of years. Whether we get it right or wrong, our ideas about likely futures really matter, so thinking about the near future is a serious task.

One important prediction we can be pretty sure about: in the next century, barring an existential catastrophe, we and planet Earth will cross a fundamental threshold as the Earth comes under conscious management for the first time in its history. In fact, we are already managing the planet's future, but so far we are doing it unsystematically and chaotically. The challenge is to manage a planet well. Have similar transitions happened around other star systems? We don't know. But what is clear is that our imaginings about the future of planet Earth are now a matter of planetary, perhaps even galactic, significance because they will shape the fate of a new complex entity — a managed or conscious planet — that is being born right now in our region of the Milky Way.

For all these reasons, *how* we imagine the near future matters. Indeed, as we saw in the last chapter, some futurists argue that studying how people *imagine* possible futures should be the main task of futures studies. As the futurist Jim Dator puts it, "futures studies does not study 'the future,' but rather . . . 'images of the future.'" But this is surely overstated. In studying imagined futures we really are trying, like Prince Arjuna, to glimpse what might actually happen. As Willis Harman, another pioneering futurist, has written, "Which futures are feasible and which are not? That is the central question of futures research."[3]

Because it matters, the one-hundred-year future is political. For the first time in human history, we face global issues including climate change, the threat of nuclear war, and new pandemics that cannot be solved nation by nation or individual by individual. Coping with them will require planetwide cooperation. After all, the craft we are hoping to steer is not the tiny raft of our individual futures, but a planet-size starship carrying billions of humans divided into two hundred distinct nations, as well as millions of other species of plants, animals, and bacteria.

Can we reasonably expect humans to build the sort of consensus that cells routinely build within multicelled organisms? We saw that macrobial cells are keen to collaborate because they are similar genetically, and specialization makes them extremely interdependent. Today, we humans are in a similar situation. We are genetically homogenous (much more so than communities of chimps, for example) and increasingly interdependent. We already see a willingness and keenness to collaborate at the level of families, communities, and even nations. Can that collaboration extend to global scales so that we humans can work together, like macrobial cells, to build a good future for the emerging macroorganism of a conscious Earth?

As we try to imagine the near future, we will ask the three basic questions of all future management: What futures do we want? What futures seem most likely? And how can we steer toward our preferred futures? We will focus on the first two steps because this is not the place for detailed programs of change. But clarity about goals and likely futures should point us in the right direction. In the early 2020s, it is clear we are not on the right track. Getting there will require large course corrections, and the great unknown is whether eight billion humans can agree to act appropriately, decisively — and in time. It will be a challenge, but there are many reasons for hope.

Step One: What Futures Do We Want?

Given the diversity of today's world, it might seem naive to expect agreement on a good future for humanity. The CEO of a major corporation, an unhoused person in a large city, a parent in a remote farming village, and a military planner will each have a different vision of Utopia. Despite this, there are good reasons to think that, as awareness of our interdependence grows, we should be able to build a broad consensus on a good future for spaceship Earth. We all belong to the same species, so we share needs, hopes, and purposes, and we are beginning to understand how tightly our fates are intertwined. Further, within networks of exchange that now embrace the entire world, billions of humans can share in global conversations about the future, and perhaps feel some loyalty to humanity as a whole.

Overlapping Visions of a Good Future

Agreeing on basic needs should not be impossible. Being able to eat enough, to play, to feel oneself part of a community, to feel free from excessive stress — these are needs shared by all humans. Indeed, we share some of them with other mammals. Who can watch dolphins or kittens at play without a flicker of fellow feeling?

One of the most influential modern discussions of a good life was initiated by Abraham Maslow's 1943 paper "A Theory of Human Motivation."[4] Maslow imagined a hierarchy of human needs, with physiological needs such as food, shelter, and health at the base; social needs such as a sense of belonging in the middle; and psychic needs such as a sense of fulfillment or meaning or "self-actualization" at the top. He believed that basic physiological needs would dominate our thoughts and behavior as long as they were unmet. But once they were met, we could try to satisfy other, "higher" needs. Maslow has been rightly criticized for focusing on needs that loom large in modern Western cultures. Nevertheless, with some corrections, the broad idea probably stands: it should be possible to agree broadly on what we understand by *a good life*.

Can we agree what sort of society might provide these "goods"? Local, national, cultural, and religious loyalties create deep divisions. Is it realistic to imagine a global consensus that can override these differences? In the Axial Age, expanding networks generated shared religious loyalties at continental scales. Can globalization generate a similar sense of shared loyalties at planetary scales?[5]

One reason for hope is the fact that different religious and ethical traditions have so much in common. In 1893, a Parliament of the World's Religions was organized as part of the Chicago World Exposition to encourage global discussion about shared ethical ideas. A century later, in 1993, a second Parliament of the World's Religions issued a "Declaration toward a Global Ethic" based on a draft by the Swiss theologian Hans Küng. It was signed by two hundred leaders from more than forty different religious traditions.[6] The declaration affirmed that "a common set of core values is found in the teachings of the religions, and that these form the basis of a global ethic." The declaration noted the unity of humanity and the interdependence of humans with one another, with other species, and with the environment: "Each of us depends on the well-being of the whole, and so we have respect for the community of living beings, for people, animals, and plants, and for the preservation of Earth, the air, water and soil." The declaration reaffirmed the Golden Rule: "We must treat others as we wish others to treat us." And it raised the metaphor of family to a planetary level: "We consider humankind our family." The principles of this "global ethic," the signatories concluded, "can be affirmed by all persons with ethical convictions, whether religiously grounded or not." In a similar spirit, the 2015 Papal Encyclical, *Laudato Si'*, talks of "dialogue with all people about our common home."

We find the same ethical overlap in the fictional imaginings of Utopian traditions. Popular Utopian visions, often expressed in revolutionary movements, focus on personal goals such as material abundance and freedom from hard labor and arbitrary oppression. In the medieval European land of Cockaigne,

"There are rivers broad and fine, Of oil, milk, honey and of wine."[7] "Big Rock Candy Mountain," a song written in the 1920s by union organizer and former hobo Harry McClintock, describes a world where handouts grow on bushes; you never have to work; the farmers' trees are full of fruit; and . . . "the little streams of alcohol come trickling down the rocks." In the early nineteenth century, a Catholic missionary described a Burmese Buddhist Utopia, in which there grew a tree named Padesa,

> on which, instead of fruit, are seen hanging precious garments of various colors, whereof the natives take whatever please them best. In like manner they need not cultivate the soil, nor sow, nor reap; neither do they fish, nor hunt; because the same tree naturally produces them an excellent kind of rice without any husk. Whenever they wish to take nourishment, they have only to place this rice upon a certain great stone, from which a flame instantly issues, dresses their food, and then goes out of itself.[8]

The Utopias of the well educated, like most elite visions of the future, were more collective. Deistic religions such as Christianity often placed their Utopias in another dimension of reality, a heaven or paradise. Secular Utopias have imagined renovated versions of Earth-bound societies, and their intent has often been satirical or critical. As we saw in chapter 3, the word *Utopia* comes from the title of Thomas More's book, which was inspired by Plato's *Republic*. The word *Utopia* is Greek and can be translated either as "no place" or as "good place." In Enlightenment Europe, a new optimism about scientific progress encouraged many thinkers to place their Utopias on Earth, and in a not-too-distant future that could be brought closer by human action.

Condorcet

One of the most interesting modern Utopias, written before the massive transformations of the past two centuries, is that of mathematician and *philosophe* the Marquis de Condorcet. His Utopia is secular; it is based on modern, scientific knowledge; and it assumes that a better world can be built here on earth. Many of his most optimistic forecasts have proved surprisingly accurate because, unlike most thinkers of his time, Condorcet glimpsed something of the astonishing technological, social, and economic transformations of the next two centuries.

Condorcet played an active role in the French Revolution until, in 1793, he was denounced by the Jacobins. In hiding, he wrote a universal history of humanity that ends with a Utopian vision of the future. Condorcet's *Sketch for a Historical Picture of the Progress of the Human Mind* was intended as a draft for a larger work, but it was left unfinished after his arrest and death in prison in March 1794. A year later, after the fall of the Jacobins, the French Convention published it in a large edition. That ensured it would have a huge influence, particularly on European thought, despite the desperate conditions under which it was written and the inevitable superficiality of some of its ideas. Condorcet's Utopian vision has roots deep in the past, because he understood the basic future-thinking principle that visions of possible futures must rest on knowledge of powerful trends in the past: "If there is to be a science for predicting the progress of the human race, for directing and hastening it, the history of the progress already achieved must be its foundation."[9]

In Condorcet's Utopia, the human race will "better itself" along three main paths. Scientific progress will raise living standards; moral progress, or improvements in "the principles of conduct or practical morality," will increase equality and respect for human rights; and medical progress will improve health, lengthen lives, and enhance people's physical and mental powers. The many areas of agreement between different religious and philosophical traditions persuaded Condorcet that building a consensus around such goals should not be hard, because all humans share a "moral constitution" based on a universal awareness of pleasure and pain and common "human sentiments."[10]

Condorcet was optimistic because he was convinced that long historical trends pointed toward increasing freedom of thought and rapid scientific and technological progress. These trends worked together in what, today, we might call positive feedback loops. Condorcet insisted that vast creativity could be unleashed by removing barriers to free thought and intellectual progress, along with the many inequalities of race, class, and gender that blocked the intellectual and moral development of so many individuals. Finally, medical progress would improve human bodies and lengthen lives: "Would it be absurd then to suppose . . . that the day will come when death will be due only to extraordinary accidents or to the decay of the vital forces, and that ultimately the average span between birth and decay will have no assignable value?"[11]

Modern Global Utopias: Balancing Growth and Limits

The fantastic scientific and technological progress since Condorcet's time has made many of his most extravagant hopes seem commonplace. But even in his time, Utopian hopes were built into some founding documents of modern political and ethical thought, such as the American Declaration of Independence and the French Declaration of the Rights of the Man and of the Citizen. Hopes for a world free of material and political oppression also appear in the founding documents of modern socialism, including the *Communist Manifesto.*

In the twentieth century, for the first time, global organizations appeared that could plausibly claim to speak for much of the world's population. Despite their political weakness, and the undemocratic nature of many governments, organizations such as the United Nations offer, for the first time in human history, a formal institutional voice for the aspirations of all humans. In 1948, the UN General Assembly adopted a *Universal Declaration of Human Rights,* the first sketch of a global Utopia to gain official acceptance on a global scale. The declaration owed something to H. G. Wells's argument, early in World War II, that if you want people to fight you must help them imagine the future they are fighting for.[12] Like Condorcet's *Sketch,* the 1948 declaration assumed that science and technological innovation would provide the material basis for a fairer and more prosperous world. We can call this the "growth" path to a better future. It assumed that, whatever the short-term costs, the long rising trends of scientific progress, new technologies, and sustained growth would eventually benefit everyone.

Since the mid-twentieth century, these hopes of a linear path to a better future have been complicated by a new awareness of planetary limits to many growth trends. That has forced us to consider "stabilization" paths to Utopia as well as growth paths.

Two hundred years ago, few thinkers expected the sort of growth we now take for granted. Most assumed that growth would soon reach limits. Condorcet, more optimistic than most, worried that rising populations could threaten progress, but he hoped that scientific and moral progress would solve the problem as people realized that they have "a duty to those who are not yet born [and] that duty is not to give them existence but to give them happiness," rather than "foolishly" encumbering the world "with useless and wretched beings."[13] Most thinkers, looking back to a past of slow technological change, were more pessimistic. Economists such as Adam Smith assumed that growth would stall once all the planet's available arable land was being farmed, while Thomas Malthus, one of history's great curmudgeons, argued that hopes of endless progress would always be stymied by resource limits. Malthus's *Essay on the Principle of Population,* first written in 1798, was a riposte to Condorcet and other Utopian writers.[14] "I have read" he wrote, "some of the speculations on the perfectibility of man and of society with great pleasure. I have been warmed and delighted with the enchanting picture which they hold forth. I ardently wish for such happy improvements. But I see great, and, to my understanding, unconquerable difficulties in the way to them." The main difficulty was how to feed growing populations because, as Malthus wrote, "population, when unchecked, increases in a geometrical ratio. Subsistence increases only in an arithmetical ratio."[15] Eventually, human populations will grow too fast for farmers to keep feeding them.

Given the slow pace of technological change for many centuries and millennia, such claims made a lot of sense. But the pessimists of Malthus's era were blindsided by the astonishing two-hundred-year boom that began in Malthus's lifetime and seemed for a time to transcend all possible limits to growth. By 1850, economic growth and technological innovation on a scale Malthus could not have imagined seemed unstoppable in industrializing countries, and a better future began to look like the inevitable outcome of a burst of technological, scientific, economic, and even "moral" progress that few had foreseen. Conflicts there might be about how to distribute increasing wealth, but, at least in the industrializing world, a better future began to seem inevitable.

In the twentieth century, however, it became apparent that the breathtaking changes of the modern era have not eliminated all limits to growth. They might have even brought them closer, as humans have begun to consume energy and resources on scales that are threatening the stability of the entire biosphere.

The first images from space helped create a new sense of our planet's isolation and fragility. As Adlai Stevenson said in 1965, "We travel together on a little spaceship, dependent on its vulnerable supplies of air and soil . . . preserved from annihilation only by the care, the work, and . . . the love, we give our fragile craft."[16]

In the mid-twentieth century, ecological warnings multiplied. One of the most controversial and influential was a book we have already met: *Limits to Growth*, published in 1972.[17] Its authors concluded, like an increasing number of environmental scholars, that building a better future would mean balancing growth with ecological restraint, and that would require decisive policy shifts and a "Copernican revolution of the mind."[18] Many growth trends, particularly in consumption of nonrenewable resources or in ecologically damaging activities such as the burning of fossil fuels, would have to be flattened or reversed in the twenty-first century in order to avoid collapse. Flattening those trends would mean abandoning hopes of endless growth. Instead, the authors argued, the world must try to preserve the gains of modernity within an "equilibrium" state in which "the basic material needs of each person on earth are satisfied" and everyone can realize their individual human potential.[19] A more stable future would not mean an end to change or to collective learning or to growth in human creativity. On the contrary, a reduced focus on economic growth might allow a greater focus on new forms of well-being. The authors cited approvingly the philosopher John Stuart Mill's description of a world less committed to endless growth:

> A stationary condition of capital and population implies no stationary state of human improvement. There would be as much scope as ever for all kinds of mental culture, and moral and social progress; as much room for improving the Art of Living, and much more

likelihood of its being improved, when minds ceased to be engrossed by the art of getting on.[20]

Growing awareness of planetary limits to growth meant designing Utopias built within those limits. The 1983 UN World Commission on Environment and Development coined the phrase "sustainable development" to refer to development that meets "the needs of the present without compromising the ability of future generations to meet their own needs." Since then, all UN declarations of this kind have imagined Utopias that balance growth and sustainability.

In the late twentieth century, global warming began to dominate discussions about limits to growth. The first international gathering to formally recognize the threat was the UN Conference on Environment and Development (the "Earth Summit"), held in Rio de Janeiro in 1992. The United Nations Framework Convention on Climate Change, which was negotiated at the Rio conference, committed UN members to "stabilization of greenhouse gas concentrations in the atmosphere at a level that would prevent dangerous anthropogenic interference with the climate system."[21] Annual Conferences of the Parties (COP) were established to assess and take global decisions about progress on climate change. The 2015 Paris Agreement, negotiated at the twenty-first COP meeting, committed to limiting global warming to two degrees Celsius and preferably no more than one and a half degrees Celsius above preindustrial levels.

In the year of the first Rio Earth Summit (1992), 1,575 world scientists, including more than 50 percent of all living Nobel laureates, issued a dire warning about human impacts on the environment:

Human beings and the natural world are on a collision course. . . . If not checked, many of our current practices put at serious risk the future that we wish for human

society and the plant and animal kingdoms, and may so
alter the living world that it will be unable to sustain life
in the manner that we know.[22]

The document concluded: "A great change in our stewardship of the earth and the life on it, is required, if vast human
misery is to be avoided and our global home on this planet is not
to be irretrievably mutilated."

Since then, the damaging trends have kept rising and so has
the number of international agreements promising serious
change. In 2000, the UN adopted eight Millennium Development
Goals, and in 2012, a second Rio conference committed to new
Sustainable Development Goals in a document called "The Future
We Want."[23] The 2015 UN statement on "Sustainable Development
Goals" is one of the clearest descriptions of a modern global Uto-
pian vision that tries to balance growth and sustainability.

We are resolved [the preamble stated] to free the human
race from the tyranny of poverty and want and to heal
and secure our planet. We are determined to take the
bold and transformative steps which are urgently needed
to shift the world on to a sustainable and resilient path.
As we embark on this collective journey, we pledge that
no one will be left behind.[24]

The document listed seventeen goals, 169 specific targets
and 232 "measures of compliance." The goals were intended as a
"blueprint to achieve a better and more sustainable future for
all," and it was hoped that most targets could be met by 2030.[25]
The seventeen main goals included the abolition of hunger,
raising basic living and educational standards for all, reducing
inequality, maintaining stable and law-abiding states, ensuring
sustainability, and supporting sustainable forms of growth.

These goals were adopted by all 193 members of the UN General Assembly in September 2015.

Such declarations skim over the many compromises that will have to be negotiated between sustainability and growth, and between the interests and goals of different regions, countries, and interest groups. But they show that, despite the endless static of political and ideological conflicts, there is emerging a broad consensus about how to build a future that preserves the gains of the modern era, while avoiding ecological overreach. Fifty years ago, such a consensus was unimaginable.

Setting targets is an important first step. But what are the chances of reaching them?

Step Two: What Futures Look Most Likely?

The second step in maneuvering our planetary ship is to find a path through the crosscutting currents and trends carrying it into the future. In today's hurricane of change, that is like piloting a ship into harbor during a storm, while arguments rage in the wheelhouse.

Trends That Point to Likely Futures

Trend hunting is the fundamental skill of modern future thinking. But it's a subtle and delicate art. The aim is to find the right balance between generality and detail. We must identify the trends most likely to shape our futures, but we must avoid overprecise predictions that are unlikely to prove accurate and could limit our options as we try to steer toward better futures.

What trends offer hints about the future of humanity? We must look for large trends with enough regularity and inertia to hint at likely global futures over decades and even centuries. We will find the most helpful trends in the "probable" and "plausible" domains of Future Cone 2 from chapter 2. Jørgen Randers, one of the authors of *The Limits to Growth*, describes how he used a similar strategy in 2012 when attempting forecasts for the year 2052: "My forecast takes as its foundation a selection of physical and ideological realities that have traditionally evolved in a sluggish manner with much inertia. . . . This rather sluggish reality constitutes what I call the 'deterministic backbone' of my forecast."[26]

Good trend hunting must also be sensitive to the many different ways in which trends, like hunted foxes, can twist and turn. Rising or "growth" trends, for example, may climb in a straight line, or accelerate and turn into exponential curves. Other trends may fluctuate like waves, or slow down, flatten, and turn into the S curves familiar in demographic history.

Expert trend hunters must also be alert to the unknowns, the unexpected surges or twists that can make nonsense of known trends. In February 2002, the US secretary of defense, Donald Rumsfeld, was asked how confident the US government was that Saddam Hussein had stockpiled weapons of mass destruction. His answer became famous: "As we know, there are known knowns; there are things we know we know. We also know there are known unknowns; that is to say we know there are some things we do not know. But there are also unknown unknowns — the ones we don't know we don't know."[27] "Known unknowns" are trends we can see, though we don't know how or when they may shoot off in new directions. Where people are involved, there are plenty of known unknowns. Sir Isaac Newton, who lost money he had invested during the South Sea Bubble in 1720, ruefully commented, "I can calculate the movement of stars, but not the madness of men."[28] "Unknown unknowns" ("unk-unks" in the jargon of techno-nerds) are the trends we can't see because we cannot even imagine them. They are what

Nassim Taleb called "black swans" because, until their eventual discovery by Dutch navigators in Western Australia in 1697, Europeans regarded black swans as impossible, mythic creatures.[29] Unknown unknowns are predictive black holes.

Of all the currents swirling around us today, which are most powerful and predictable? Which are pointing toward our Utopias? Which should we avoid? And what unknowns could send us lurching in unexpected new directions?

The big-history perspective encourages us to keep an eye on very large trends. One large trend is peculiarly interesting because it appears at all scales and in many different domains. This is the pattern that the evolutionary biologists Niles Eldredge and Stephen Jay Gould called "punctuated equilibria."[30] They argued that in evolutionary biology the sort of steady gradual change that Darwin thought of as the evolutionary norm is in fact exceptional. Most new species appear unexpectedly, and then their populations rise rapidly at evolutionary "punctuations" until they reach a sort of demographic plateau as they settle into their new niche. Then populations fluctuate for thousands, perhaps even millions of years before eventually they fall and the species goes extinct. It turns out that similar "tabletop mountain" patterns can be found well beyond biology because they describe the histories of all complex entities, from molecules to mole rats, from stars to you and me. All complex structures emerge, often rapidly, at "threshold" moments, before settling into a relatively stable equilibrium and eventually breaking down.[31] This pattern is ubiquitous because it arises from a tension between two other universal trends: the rising trends that allow the emergence of complex entities (a trend for which, curiously, we have no general scientific label); and the falling trends, governed by the second law of thermodynamics, that eventually break down all forms of complexity.

We will see that the pattern of punctuated equilibria may provide important clues about the future of humanity and planet Earth. But if we narrow our focus to the scale of human history (a mere two hundred thousand to three hundred thousand years!), we mainly see long rising trends, driven by collective learning. Some have persisted for much of human history, as humans have multiplied and colonized and exploited our surroundings with increasing ingenuity. In the last two centuries, some of these trends have accelerated spectacularly, creating a modern world in which growth seems normal. But we now know that some rising trends such as population growth and consumption of energy and resources are reaching planetary limits and starting to flatten. And that is just what you expect in the history of all complex entities, when the emergent or "growth" phase of punctuated equilibria bends toward a more stable plateau-like phase. Taken together, the long rising trends of human history, the newer stabilizing trends emerging as we reach planetary limits, and the universal trend of punctuated equilibria hint at the possibility that something new is emerging on planet Earth.

We will follow these hints as we look more closely at the three types of trends that will shape the next century: growth trends, stabilizing trends, and the erratic and unpredictable trends of politics: the fights, discussions, and negotiations in the ship's wheelhouse that will eventually determine the course our planetary vessel takes over the next few decades.

Growth Trends and the Future

On the scale of human history, there are many powerful rising trends driven by collective learning. Populations have grown; human technologies have become more powerful; human exchange networks have expanded; human consumption of resources has risen. In recent centuries, some of these trends have accelerated so sharply that they have begun to look exponential.

Energy use demonstrates how spectacular some of these trends are. A single human can generate about 150 watts (1 watt is a flow of 1 joule per second), or one-fifth of a horsepower (1 horsepower = about 735 watts).[32] Ancient technologies such as the use of fire or the taming of horses, camel, and oxen increased the average power available to each person to about one horsepower. In the last two centuries, the energy at our disposal has increased many times over. Fossil fuels are being used on such a scale that, on average, each person now controls one hundred horsepower, or 73,500 watts. Those in charge of machines such as airliners routinely control many millions of watts. These figures offer a crude measure of how collective learning has increased the power of our species and how that increase has accelerated in the modern era.

As Condorcet argued, many of these growth trends seem to point to a better future for humanity. Growth in scientific and medical knowledge and growth in wealth and education count as generally benign forms of growth. They have improved lives. They have brought elements of Condorcet's Utopia to much of the world, first in Europe and the North Atlantic region, and then, with a lag of about a century, to most other parts of the world. Economic growth is an ancient trend whose significance has been transformed in the modern era. For most of history, gains from economic growth did little to raise the living standards of most people. They were gobbled up by population growth or the expenses of governments and elites, so that most people continued to live close to subsistence, never far from the threat of starvation. Even in 1800, more than 80 percent of the world's population lived below the modern international poverty line of about two or three dollars a day. Then things changed, fast. By 2017 fewer than 10 percent lived below that level, despite rapid population growth and the increasing wealth of elites.[33] The growing wealth of nonelites is new in human history and counts as one of the great achievements of the fossil fuels revolution. In the same period, average life expectancies more than doubled, rising from about thirty years to more than seventy years, while the percentage of children dying before the age of five fell from more than 40 percent to about 4 percent. As the scientist Vaclav Smil has argued, infant mortality may be the best single indicator of improving quality of life because so many medical, sanitary, economic, and social structures need to change if more babies are to survive.[34] Modern machines have eliminated much grinding physical toil; adult illiteracy has fallen from about 88 percent to about 14 percent; and modern medicine has almost eliminated diseases such as smallpox and polio, and greatly reduced suffering from other diseases. No one should underestimate the importance of anesthetics, first introduced in the middle of the nineteenth century. Fanny Burney's description of a mastectomy without anesthetics in 1810 is a vivid reminder of the horror of operations before anesthetics.[35] Today, for the first time in human history, a tooth can be pulled, or a limb amputated, without agony and the near certainty of infection.

Less clear, but striking nonetheless, are growth trends in what Condorcet called "moral progress." We have seen hints of a slow growth in the circle of concern, both at individual and governmental levels. On paper, at least, most modern governments profess respect for the basic human rights of everyone. That is new. So are the modern ideas that slavery and gender and racial inequality, institutions regarded as normal for thousands of years, are unacceptable. Furthermore, despite the terrible stories that dominate press coverage and social media (because "if it bleeds it leads"), levels of interpersonal violence, including judicial torture, have fallen sharply in the modern era, and casual violence is increasingly unacceptable.[36] Though the practice of many modern communities falls short of their formal commitments, there are many signs that better material living standards, by freeing people from the constant desperate fight for scarce resources, have led to a softening of manners that Condorcet would have recognized as "moral progress."

Technological innovation drove the first stages of another growth trend that Condorcet missed: the first cautious steps into space. Yuri Gagarin was the first human to enter space when, in 1961, he made a single orbit of Earth. In 1969, Neil Armstrong became the first human to set foot on another planetary body, the moon. Early in the twenty-first century, space robots have visited several planets and moons of our solar system, and the two Voyager satellites, launched in 1977, have reached the outer edges of our system. In 2021, the first tourists flew in space, and several nations are planning manned expeditions to the moon, to nearby asteroids, and to Mars.

Planetary Limits and Stabilizing Trends

Collectively, humanity has gained much from growth. But despite these gains we now know that many rising growth trends must be reined in.

Some growth trends are flattening spontaneously. Global rates of economic growth (a loose proxy for human consumption of resources) have slowed from about 5.5 percent per annum in the decade after 1961 to just over 2 percent per annum fifty years later in the decade after 2011, despite rising fast in regions such as China and India.[37] In retrospect, the remarkable growth rates of global GDP in the mid-twentieth century look like the beginning of a slowdown.

More spectacular, but similar in its timing, is the slowdown in population growth after two centuries of exceptionally rapid growth. Rates of growth began to slow from the late 1960s, and what had begun to look like an exponential rising trend started bending into an S curve. The first evidence of a slowdown appeared in 1968. Coincidentally, that was the year in which two modern Malthusians, Paul and Ann Ehrlich, published a bestseller called *The Population Bomb,* warning of imminent global collapse due to overpopulation.[38] Their bad timing should remind all would-be futurists how easy it is to get things spectacularly wrong! Today, most demographers expect human populations to peak later this century, at a number between nine and twelve billion, before starting a long, slow decline.[39] The change is momentous because the long trend of population growth has persisted, with temporary ups and downs, for most of the past ten thousand years. That long growth trend seems to be flattening. Slowing population growth will reduce pressure on global resources and environments, but it will also slow economic growth as the number of wage earners stabilizes and populations age.

The Ehrlichs were wrong-footed, like Malthus, but for different reasons. Population growth began to slow not because we ran out of food and resources, but because of changing demographic behavior driven by changing lifestyles. In the agrarian era, most people were peasants, and the best way of building wealth was to have as many children as possible who could work your farm and support you in old age. Yet up to half of all babies died in childhood. So, to maximize the number of children who reached adulthood, women had to bear as many children as possible. That is a major reason why childbearing and child-rearing dominated most women's lives throughout the agrarian era.

The modern era changed these ancient demographic calculations, as we became a species of city dwellers and wage earners. In modern towns and cities, more children survive to adulthood because of better access to wages, food, and healthcare. But rearing urban children costs more because parents must pay for food and education, and urban children cannot usually be put to work as early as those in farming villages. For these and other reasons, most modern urban families prefer to have fewer, but healthier and better educated, children. From an era of high birth and death rates, we have entered an era of low birth and death rates, which is transforming the lifeways of women throughout the world, by removing a crucial driver of gender inequality. In 1800 women had, on average, 5.8 children; in 1950 the average was still about 4.8; but then it fell fast to 2.5 in 2014.[40] Populations grew explosively in the past two centuries because death rates began to fall well before birth rates, as a result of improvements in food production and healthcare. Birth rates began falling in wealthier and more urbanized countries in the nineteenth century, but not until the late twentieth century did they begin to fall globally, eventually catching up with falling death rates in the 1960s.

Rates of population growth are falling without massive political intervention, though the slowdown could be accelerated by systematic government action, such as improving the educational and professional opportunities available to young girls. Other dangerous growth trends will flatten only after difficult, deliberate, and large-scale political interventions.

One of the most dangerous growth trends tracks human impacts on the global climate system. Human technological creativity has always put pressure on local environments, but until the modern era, few imagined that human activities could alter environments on planetary scales. One of the first to glimpse the scale of modern human impacts was the Swedish chemist Svante Arrhenius. In the 1890s, he calculated that humans might be burning enough fossil fuels to warm the Earth's atmosphere through what came to be known as the "greenhouse effect."[41] In the 1960s, in Hawaii, the atmospheric scientist Charles Keeling began measuring levels of atmospheric carbon dioxide and found they were rising fast. The so-called Keeling curve has kept climbing, and as Arrhenius had anticipated, average global temperatures are climbing alongside it. In 2021, they are about one degree Celsius above preindustrial levels, and there is increasing evidence that rising temperatures are triggering anomalous climatic events, such as extreme storms, floods, and fires. Long-term studies of climate change, based on the analysis of air bubbles trapped in ice cores, show that over the past two hundred years carbon dioxide levels have risen far above the highest levels seen in the past million years. No wonder. Since 1800, carbon dioxide emissions have increased by more than a thousand times.

The most influential long-term climate forecasts today are those of the UN's Intergovernmental Panel on Climate Change, or IPCC, which was created in 1988 by the United Nations and the World Meteorological Organization. Since 1990, the IPCC has produced six reports that summarize research by hundreds of scientists from all parts of the world. The first part of the sixth report, published in August 2021, includes five possible scenarios ("Shared Socioeconomic Pathways" or SSPs) for climate change based on already existing climate change policies and technologies.[42] The IPCC reports represent modern future thinking at its most sophisticated. They model likely trends in global climates with a rich understanding of the causes of climate change and the associated probabilities and uncertainties; they use colossal amounts of information from research by scientists in many different countries; and they analyze that information with all the computational power of modern supercomputers. It would be very foolish to ignore these forecasts, even though we know that they, like all forecasts, could be turned upside down by unknown and unimagined trends and events.

The 2021 IPCC report finds that, under all of its five SSP scenarios, global temperatures are likely to reach one and a half degrees Celsius above the level for 1850–1900 by 2040. Only under its most optimistic scenarios will global temperatures be held to less than two degrees Celsius above that level by 2100, and under the least optimistic scenarios, they could exceed that level by more than five degrees Celsius. Average global temperatures of more than two and a half to three degrees Celsius above the 1850–1900 levels have never been experienced in the course of human history.[43] Such temperatures will drown coastal cities and expand deserts; they will trigger more erratic and violent climatic shifts, more climatic swings, more droughts, more floods, more wildfires, more cyclones; and they will undermine food production and accelerate the spread of new diseases. Like prodding a caged animal, rising greenhouse gas emissions may provoke violent ecological tantrums. For example, vast amounts of methane, a more potent greenhouse gas than carbon dioxide, are frozen in the oceans in so-called methane clathrates. As oceans warm, methane clathrates will eventually melt, suddenly releasing lots of methane into the atmosphere. That could trigger other tipping points such as a switch in the North Atlantic Current that might bring Arctic climates to much of Europe. (The 2021 IPCC report concludes that the North Atlantic Current is "very likely" to weaken in the twenty-first century, but there is "medium confidence" that it will not collapse before 2100.)[44] Tipping points are known unknowns, like the illustrations of unknown lands on medieval European maps. They tell us that "beyond two degrees Celsius warming there be monsters!" Tipping-point monsters could make things ugly for our grandchildren and their children.

Carbon dioxide stays in the atmosphere for a long time, so global warming will continue long after greenhouse gas emissions start falling. But drastic cuts to emissions now could limit warming to less than two degrees Celsius above preindustrial levels by 2100.

Another dangerous trend affects the diversity of species on planet Earth. Humans and their domesticates are consuming more and more resources and energy, and other species are suffering. By 2020, humans and their animal domesticates made up more than twenty times the combined biomass of all other land mammals, while the biomass of domesticated chickens was more than twice that of all other bird species.[45] The 2019 IPBES

(Intergovernmental Science-Policy Platform on Biodiversity and Ecosystem Services) report found that the rate of species extinction was "at least tens to hundreds of times higher than the average rate over the past 10 million years and is accelerating."[46] This is tragic. But it is also dangerous for us humans because our survival and well-being depend on the organisms all around us, from trees to fish to pollinating insects like bees.[47] Can our descendants hope to flourish in an ecologically impoverished biosphere?

Messing with environments could trigger unknown new trends. We live, now, in novel biochemical environments created over the past two centuries as humans have manufactured millions of new materials and chemicals from plastics and fertilizers to concrete and mind-altering drugs. We don't really understand what this may mean for the biosphere as a whole. In *Silent Spring*, published in 1962, the biologist Rachel Carson documented the terrifying impact of pesticides and other agricultural chemicals on humans and other species. The 2020 pandemic reminded us of something epidemiologists have been saying for years: we are transforming our epidemiological environments in ways that help diseases proliferate and move between species in new ways. In modern humans, viruses have found a mode of transport that can carry them around the world in just days.

We now know that the planetwide geological, biological, and atmospheric system that the scientist James Lovelock once called "Gaia" is extraordinarily complex, but it is also bounded.[48] Over fifty years, the idea of planetary limits to growth has worked its way into most serious thinking about the near future. In answer to the question: "How far can we push planetary systems before they break down?" a team of scholars headed by the climate scientist Johan Rockström has defined several "planetary boundaries," beyond which the global environment could start to collapse. The boundaries include high levels of atmospheric greenhouse gases, ozone-depleting chemicals and aerosols, and ocean acidification, as well as deforestation, declining levels of biodiversity, and declining freshwater reserves.[49] Their research suggests that we have already crossed some boundaries, including those for biodiversity and nitrogen flows. As Rockström writes, "We are the first generation to know that we're undermining the ability of the Earth system to support human developments."[50]

Another growth trend that we must flatten tracks the destructive power of human weapons. In the nineteenth century, cannons were the most powerful weapons of war. At their deadliest, they could kill quite a few enemy soldiers. Two centuries later, we have weapons that are vastly more destructive. The fission bomb dropped on Hiroshima on August 6, 1945, killed at least sixty thousand people almost instantly, and a similar number died of radiation and injuries in the years that followed. In the next few decades, the United States and the USSR developed much more powerful fusion bombs, and the number and power of nuclear weapons in their arsenals increased to grotesque levels. By 1980, almost one hundred thousand nuclear warheads were deployed. If unleashed, they could have ruined the biosphere in just a day or two. The main destroyer would not have been the bombs themselves, but the dust clouds, which would have been so high in the atmosphere that they could not be rained out, blocking sunlight for months or years. That would have created a nuclear winter, darkening the skies and ruining agriculture around the world. We have come very close to nuclear catastrophe. Just after the 1962 Cuban Missile Crisis, President Kennedy said the odds of all-out war had been between 1 in 2 and 1 in 3.[51] Since then there have been other near misses that have taught us the meaning of MAD, or mutually assured destruction, an acronym coined at Herman Kahn's Hudson Institute in 1962.

More through luck than skill, we have avoided a nuclear Armageddon so far, and the number of nuclear warheads began to fall in the late 1980s. But in 2021 nine countries were thought to have about thirteen thousand nuclear weapons, about fifteen hundred on hair-trigger alert, which means they are ready to be launched. And the number of nuclear weapons is rising once more. The threat of sudden collapse through nuclear war will hover over all attempts to build a better world until some way is found of persuading those who own such weapons to destroy them along with the means of building them.

One more growth trend that may threaten human futures is rising inequality. Systematic inequalities based on multigenerational differences in wealth and coercive power emerged in the agrarian era as agricultural societies produced surplus resources that could be distributed unevenly. For much of the last five thousand years, small elites controlled most surplus wealth, while most people lived as peasants close to subsistence. Such inequalities create social tensions and, if taken to extremes, they can lead to social breakdown. Allowing inequality to keep rising is like stretching a spring and hoping it doesn't snap.

In the modern era, growing wealth, changing lifeways, and human rights movements have done much to soften inequalities of race and gender, but fundamental inequalities in wealth and power have, if anything, increased. In the late nineteenth and early twentieth centuries, global inequalities took extreme forms as those nations that industrialized first, most of them in Europe or the West, used their wealth and their technological and military power to dominate much of the rest of the world. In the second half of the twentieth century, decolonization and the take-up of modern technologies in other regions narrowed some of these inequalities. But deep inequalities remain, and they continue to fuel mass migrations and local wars. The impacts of climate change will be felt most acutely in countries that contributed the least to global warming and that have the least capacity to respond to it. And the jockeying for advantage between powerful nations that drove so many twentieth-century wars will persist, fueled by modern inequalities and by memories of the deep inequalities of the imperialist era, as rising powers such as China and India challenge former hegemons in Europe and the United States.

Inequalities *within* nations drove many of the revolutions and civil wars of the modern era. Socialists expected such inequalities to bring down the wealthiest "capitalist" societies, but curiously, they did not. As the French economist Thomas Piketty has shown, for much of the twentieth century, levels of inequality within nations declined, particularly in wealthier capitalist societies.[52] This was partly because the world wars of the early twentieth century destroyed traditional forms of landed wealth, while increasing production allowed governments in industrializing countries to buy off discontent by building welfare systems to protect the less wealthy. Finally, capitalist entrepreneurs and governments discovered that it was possible both to increase profits and reduce social tensions by raising the wages and living standards of wage earners, because wealthier workers were less discontented and could buy more products. For these and other reasons, living standards rose spectacularly in the West in the twentieth century and then in many other countries, including Asia during the spectacular economic booms of the Chinese, Indian, and other "tiger" economies beginning in the 1980s.

From the 1970s, levels of inequality within nations began to rise again, first in the capitalist West and then in much of the rest of the world. Governments of wealthy capitalist countries, under the guidance of neoliberal free market economists, began to dismantle the redistributive mechanisms created earlier in the century, arguing that they limited the free pursuit of profits, reducing growth rates that should, in theory, benefit everyone. After the collapse of the Soviet command economy, inequalities rose sharply in former Soviet bloc countries, and rapid industrialization in countries such as China and India also increased inequality. Today, in many countries, inequality is returning to the high levels of the late nineteenth century. In 2018, it was estimated that slightly more than half the world's private wealth was held by 1 percent of the global population.[53]

The rising trends of inequality, like those of climate change, can cross unpredictable tipping points. Who can predict when inequality will trigger revolutionary breakdown? What makes these tipping points particularly dangerous today is the existence of nuclear weapons that could turn a class war into a global Armageddon. Limiting extreme inequalities will not be easy. In a recent global history of inequality, the historian Walter Scheidel argues that inequality has never fallen significantly through deliberate human efforts. It has fallen only as a result of catastrophes such as wars, state breakdowns, natural disasters, or pandemics.[54] We must hope that this rule can at least be softened in the future. And it may be that in a more stable future world less committed to limitless economic growth, the challenge of maintaining political and social stability by reducing inequality will be taken more seriously than it is today.

Unknowns: The Politics of the Future

Many of the trends we have looked at are regular enough to allow some plausible forecasts. They lie in the "probable" or "plausible" domains of Future Cone 2, and we understand many of them much better than we did fifty years ago. But how we respond to these trends will depend on politics. And most political processes are too irregular to predict with confidence. They mostly lie in the "possible" domain of predictability. Will humans forge a consensus strong enough to allow the coherent, intelligent, and decisive action needed to build a sustainable

Utopia? Or will squabbles over interests, goals, and costs prevent the needed course corrections?

No confident predictions can be made because in politics there are few regular trends and many unknowns. Hopes for a better world could be sabotaged by small groups armed with nuclear weapons or bioengineered viruses, or by bad science, or political procrastination, or leaders with too narrow a vision of the future. Or by the sheer complexity of the problems we face. But there are some promising if fragile long trends in political history. Three are particularly hopeful, and we have already seen them. The first is the emergence of planetwide human networks of exchange, which is increasing awareness of shared global challenges and beginning to create delicate new loyalties and commitments to the global community. The second trend is the emergence of global coordinating institutions such as the United Nations and the many NGOs that work, today, across national borders. They can give a voice and some political weight to emerging global concerns, and many of these institutions share the same broad vision of a better future. Global businesses can also play a crucial role in building sustainable societies once it is accepted that capitalism can flourish only in an ecologically sustainable world. The third trend is what Condorcet might have called "scientific progress." We know so much more about the planetary system than we did even a few decades ago, we have many of the technologies we will need to build a sustainable future, and more are on the drawing board.[55]

These are promising trends, but not guarantees. At the twenty-sixth Conference of the Parties to the UN Framework Convention on Climate Change, in Glasgow in November 2021, UN member states made a series of commitments that almost certainly fell short of those needed to keep warming below one and a half degrees Celsius above preindustrial levels by 2100. But they did suggest an increasing sense of global urgency about the issue. Will countries follow through on these commitments?

Imagining Possible Future Scenarios

The trends we have looked at offer tantalizing hints about what may happen in the next hundred years. We have identified powerful growth trends, but many of those trends may now be flattening. That is a pattern we have seen before. A turn toward more stable trends is what you see just after a star lights up, or when, after a period of rapid growth, a new biological species settles into a new phase of maturity. It is what you see when a new complex entity emerges, and now we can glimpse what it is that is being born here on planet Earth: a conscious planet.

An ancient complex entity, planet Earth and its cargo of life, is being transformed by what is, on planetary scales, a sudden mutation: the appearance of a species that has acquired enough power to shape the future of the biosphere through conscious activity. The sphere of human thought, which the great Russian geologist Vladimir Vernadsky called the noosphere, has suddenly become powerful enough to change Earth's future.[56] Earth is becoming conscious in the sense that our bodies are conscious: most processes will continue as in the past without conscious thought, but from now on many big decisions about planetary futures will depend on conscious decisions taken by humans. The change is already under way because we are already transforming planet Earth. The only issue is how well we will manage the transition.

We can imagine many ways in which the transition to a conscious planet might occur, and not all assume a successful outcome. It can be helpful to imagine several scenarios, all aiming at the forecaster's sweet spot between precision and generality, and distributed along a sort of normal curve under which some are more probable than others. The good news is that in all but the most catastrophic future scenarios, we humans will probably succeed, perhaps after a long apprenticeship, in becoming competent, and perhaps reasonably good planetary managers. We already know many of the things we must do; we understand the planetary system much better than we did even fifty years ago; there have been profound shifts in popular and governmental attitudes to the environment; and we have most of the tools and resources needed to succeed. As Jørgen Randers writes, "100% wind, water, and solar energy can be achieved with existing technology. Nor is lack of money the real issue. War spending is over 2% to 3% of world GDP. It would take much less than this to cover the cost of bringing greenhouse gas emissions down by 50% in twenty years and do the necessary adaptation to residual climate-change impacts."[57]

We also have the political and economic levers needed to act fast and decisively. The radical transformation of the US and Soviet economies within a year of being attacked in World War II shows how fast modern states can change direction when the political debates are over. Equally striking are the policy U-turns made by most governments to cope with the COVID-19 pandemic. Once there is a global consensus, change could come fast.

Four General Scenarios for the Near Future

Building on the ideas and trends described in this chapter, we can start imagining more specific pathways to alternative futures. These will range more widely than the Shared Socio-economic Pathways (SSPs) of the IPCC's sixth report because, unlike that report, they allow the possibility of fundamental policy shifts in the near future.

From now on I will dispense with the qualifications and ifs and buts of careful forecasting, and simply describe some possible futures suggested by the discussion so far. Some of these scenarios may look quaint quite soon; some will look over-optimistic, others will look too pessimistic, and on an imagined normal curve of possible futures, some are more likely than others. Nevertheless, imagining possible futures is itself a crucial tool of future thinking. These speculative scenarios are offered in a spirit that is both playful and serious; playful because we always play with possible futures in our imagination, and serious because we do care which imagined futures will eventually turn up.

The four global scenarios described below are adapted from scenarios developed by the futurist Jim Dator, who worked for many years with groups collectively imagining possible futures.[58] He found that most imagined futures fall within four categories that he calls "collapse," "discipline," "transformation," and "continued economic growth" ("business-as-usual" is an alternative label). Versions of Dator's taxonomy have been used by many futurists. Dator resists ranking different scenarios, but we will relax this rule and assess possible futures according to their closeness to the global Utopias described earlier in this chapter. Doing this is important because, as H. G. Wells argued, credible images of better futures can create hope, and hope is itself a powerful motivator. My own four scenarios, though recognizably descendants of Dator's, are labeled "collapse," "downsizing," "sustainability," and "growth."

Whatever scenario eventually materializes, the path to it will be turbulent. Technological and economic turbulence is guaranteed by the sheer complexity of the shift from fossil fuels to more sustainable technologies, and the certainty that many mistakes will be made along the way. Political turbulence is guaranteed by conflicts between the needs of the emerging global system and those of regions, states, and other localized interest groups. The United States, the dominant global power of the late twentieth century, is being challenged by emerging rivals. The democratic capitalist states that dominated the world after the collapse of the Soviet Union have lost confidence, partly because slowing growth and increasing inequality have sapped the optimism and reduced the size of their once prosperous middle classes. Meanwhile, rising powers such as India and China have become more confident and assertive, and modern military technologies have turned small dissident groups into serious threats to global peace. Dangerous conflicts are likely. But we should also be on the lookout for the crucial political tipping point at which leaders in many parts of the world, persuaded perhaps by regional climate-driven catastrophes, agree that their own vital interests require serious global collaboration on the task of building sustainable futures.

Only under extreme "collapse" scenarios will we completely fail our audition as planet managers. In the worst scenarios, human societies will be brought down by some combination of starvation, warfare, political and economic collapse, and pandemics. Our species may be driven to extinction. With humans out of the way, the biosphere will repair itself over many centuries. Toby Ord has made an ambitious attempt to estimate the probability of "existential catastrophes," catastrophes that destroy humanity's "longterm potential," blocking possibilities of a good future for thousands of generations.[59] On his very tentative estimates, about 16 percent (one-sixth) of possible futures lead to an "existential catastrophe." Of course, these estimates should not be taken too seriously, but they may point to the right orders of magnitude. What stands out from Ord's estimates, summarized in table 8.1, is that the most dangerous threats to our future arise from human technological and economic overreach. That is actually good news because it means that, in principle, humans should be able to manage those threats.

Table 8.1. The Likelihood of Different Existential Crises

Existential catastrophe via . . .	Chance within the next 100 years
Asteroid or comet impact	~ 1 in 1,000,000
Supervolcanic eruption	~ 1 in 10,000
Stellar explosion	~ 1 in 1,000,000,000
Total Natural Risk	~ 1 in 10,000
Nuclear war	~ 1 in 1,000
Climate change	~ 1 in 1,000
Other environmental damage	~ 1 in 1,000
"Naturally" arising pandemics	~ 1 in 10,000
Engineered pandemics	~ 1 in 30
Unaligned artificial intelligence*	~ 1 in 10
Unforeseen anthropogenic risks	~ 1 in 30
Other anthropogenic risks	~ 1 in 50
Total Anthropogenic Risk	**~ 1 in 6**
Total Existential Risk	**~ 1 in 6**

(Based on Toby Ord, *The Precipice,* p. 167.)

Toby Ord's notes: "My best estimates for the chance of an existential catastrophe from each of these sources occurring at some point in the next 100 years (when the catastrophe has delayed effects, like climate change, I'm talking about the point of no return coming within 100 years). There is significant uncertainty remaining in these estimates and they should be treated as representing the right order of magnitude — each could easily be a factor of 3 higher or lower. Note that the numbers don't quite add up: both because doing so would create a false feeling of precision and for subtle reasons [discussed elsewhere]."

*Unaligned to human interests.

Less dire scenarios can be found in the remaining 84 percent of Ord's imagined sample space of possible futures: those in which humans survive. They include less extreme collapse scenarios that are not fatal to our long-term survival, but pretty catastrophic all the same. Under these scenarios, societies will fail to grapple with the complex challenge of managing a biosphere, and human societies will enter a dark age that could last for centuries.[60] Wars will devastate whole regions, pandemics and famines will kill millions, most survivors will eke out a subsistence living, and wealthy elites will live in fortified special settlements. Most of the gains of the modern era will be lost, including gains in human rights; slavery and harsh forms of gender and racial inequality could return along with the routine use of corporal punishment, mutilation, and torture, as shortages drive increasingly desperate battles for survival. With less food, declining medical care, and limited supplies of anesthetics and basic medicines, life expectancy may return to premodern levels. Looking back in a few centuries' time, most of those alive today will look like a privileged precrisis cohort of humans, the last to enjoy the spoils of the fossil fuels revolution. Our descendants will wonder what went wrong and why we let things fall apart.

In "downsizing" scenarios, governments and societies, having concluded that growth itself is the main problem, will focus on sustainability goals and aggressively rein in many growth trends. Governments will slow growth with heavy taxation and direct regulation of unsustainable activities. That will mean a more spartan world with few luxuries and strict limits on family size. Average material living standards for most people will be lower than those of the early twenty-first century, though wealthy elites may enjoy higher living standards in protected bubbles. Liberal democracies will struggle in such an environment, because it will be hard to build consensus when resources are shrinking.[61] So downsizing societies will often use authoritarian methods. This is a scenario that respects stabilization goals but is less respectful of the freedom and equality goals of modern Utopian declarations. Our descendants will have learned how to manage a planet sustainably but may have lost many political and legal rights. Scientific and technological innovation will continue, though perhaps less rapidly than in less authoritarian scenarios. At their worst, downsizing futures may look dystopian, even Orwellian, though at their best they may look like the anarchist world of Anarres from Ursula Le Guin's *The Dispossessed.* Space exploration will continue, possibly on a large scale if backed by governments competing for the resources of asteroids, moons, and planets.

"Sustainability" scenarios combine the optimistic spirit of Condorcet with realism about planetary limits. They envisage futures in which high living standards are maintained by sustainable technologies but hopes for never-ending increases in consumption have been abandoned. Sustainability scenarios come closest to the Utopian goals described earlier in this chapter. In *Journey to Earthland,* Paul Raskin imagines a turbulent few decades as traditional governments with traditional growth values compete with the sustainability values of a Global Citizens Movement. By the middle of the century, as the dangers of ecological and political collapse become more apparent, a broad global consensus emerges on the importance of sustainability, along with new global institutions to coordinate global action. The values of most societies shift toward sustainability, increasing equality, and improved quality of life, rather than increasing consumption.[62] In sustainability scenarios, the idea of global citizenship will become as self-evident as national citizenship is today. But regional and cultural diversity should continue to flourish as they do within many multicultural nation-states today.

Technological innovation will continue, driven by the most powerful of all drivers of human history, collective learning.

Innovations will include new ways of generating energy sustainably, cheaper and more efficient forms of transportation and manufacturing, new ways of mopping up or burying greenhouse gases, and many types of smart robots. They will also include medical interventions that can slow aging (some physicians are already treating aging as a disease, not an inevitability), extend life spans, provide new smart prosthetics, and cure diseases such as cancers.[63] Sustained innovation and drastic reduction in waste, including military expenditure and advertising, will ensure living standards higher than those of the early twenty-first century. But increasing awareness of the dangers of overconsumption will encourage an ethos of sufficiency and broad equality. The assumption that progress means neverending growth will be replaced by new definitions under which progress of many kinds continues within a stable equilibrium state that keeps growth and population within what the authors of *Limits to Growth* called a "carefully controlled balance."[64] Global populations will stabilize at eight billion or less, universal healthcare and education and guaranteed material security will begin to seem normal, and most people will work fewer than twenty hours a week. They will be "time affluent" and will live with new and more ecologically realistic ideas of what is meant by a "good life." In such scenarios, humans will eventually learn how to manage the biosphere well, in a spirit of stewardship. National governments will survive, but the power of global governmental organizations will increase, and their authority will be accepted by most communities on Earth.

"Growth" scenarios downplay the significance of limits to growth and envisage a future dominated by the large growth trends of the modern era. In these scenarios, economic growth and innovation will be powered by the capitalist alliance of governments and businesses that has driven so many growth trends in the modern era. Supporters of growth scenarios are techno-optimists. They reject the need for fundamental course corrections on the grounds that a flourishing capitalism will spawn the technologies needed to combine growth and sustainability, while governments will step in with large eco-engineering projects where necessary. In the most optimistic growth scenarios, sustained economic growth and new technologies will keep raising consumption levels and material living standards for most people, while solving the more dangerous ecological challenges. Businesses will discover that profits can be made from sustainable technologies. Levels of inequality will surely rise in highly competitive economic environments, creating significant social instability. As in most other scenarios, some will experiment with the biological and genetic modification of human beings. That will bring medical benefits but may also create new social divisions as small populations of genetically or bionically engineered humans begin to appear, mainly among the rich who can afford costly medical interventions.[65] Some people will live to 150 years or more; enhancements to hearing, eyesight, and intelligence will become commonplace, and prosthetic limbs will be controlled directly by the brain or perhaps simply regrown. Human population growth will slow, though not as fast as under sustainable or downsizing scenarios.

If advocates of continued economic growth scenarios are wrong, as the *Limits to Growth* literature argued, growth scenarios may turn into collapse scenarios. We may cross tipping points in global warming, systems of food production may break down, and conflicts over resources may lead to catastrophic warfare.

Even in the three more optimistic scenarios, the past will weigh heavily on the future. Existential threats such as nuclear war or engineered pandemics will hang over humanity and the world will be uncomfortably warm for centuries. Many of today's cities will be under water, ocean acidification will have reduced fisheries and marine biodiversity, deserts will have spread, and weather events that seem extreme today will seem commonplace. Extinction rates of other species may accelerate under extreme growth scenarios but will slow in sustainability and downsizing scenarios. However, in all scenarios the world will be biologically impoverished even by today's standards. Competition will lead to wars, but only in collapse scenarios will war bring down the global system. There will be considerable inequality in all scenarios, though it will be limited under sustainable and downsizing scenarios. The ancient driver of collective learning will ensure that technological, scientific, and artistic creativity continues in most scenarios, and there will also be creativity in social organization. New scientific paradigms will transform our understanding of phenomena such as dark matter or the relationship between relativity theory and quantum physics or perhaps even consciousness. We will learn if there is life beyond Earth. Finally, by the end of this century, in most future scenarios, there will be small, pioneering colonies of humans throughout the solar system, and robotic spacecraft will have set off for other star systems.

Figure 8.1 (on the following page) imagines changes in human population levels under different scenarios in order to highlight some of the key differences. The details should not be taken too seriously, but the general shape of the trend lines may capture significant differences between these scenarios. In practice, of course, none of these scenarios is likely to play out in pure form. The future that turns up will be a complex and contradictory mash-up of many of the trends we have seen, both good and bad, spiced with trends and events we cannot imagine today. When futures arrive, they will surely be as messy as the past.

Step Three: What Action Needs to Be Taken?

This is not the place for a detailed plan of action. But action is needed if we are to avoid collapse and steer toward the more optimistic scenarios. The trends we have looked at tell us where

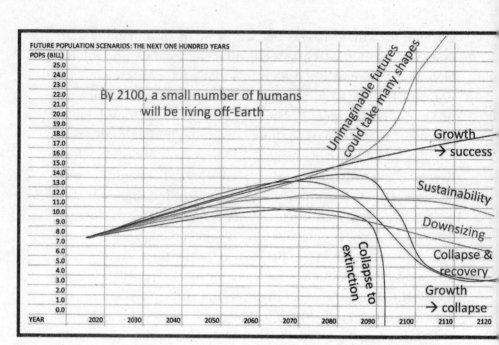

Figure 8.1: Populations up to 2120 under Four Different Future Scenarios: Growth, Sustainability, Downsizing, and Collapse

we should be heading. "The main solution is so simple," Greta Thunberg told the World Economic Forum in 2019, "that even a small child can understand it. We have to stop the emissions of greenhouse gases. And either we do that, or we don't."[66] In all the more optimistic scenarios, we will have to become competent planetary managers. That will mean building economies in which prices better reflect the real environmental costs of goods and services, and greenhouse gas emissions are cut drastically. Even in growth scenarios it may mean a slowdown in rates of economic growth. We must slow or reduce the extinction rates of other species and slow or reverse growth in the number of humans and human domesticates. To avoid catastrophic conflicts, we will have to control dangerous weapons and limit extreme inequality. The 2015 Paris Accord, the UN Sustainable

Goals, and many other global treaties provide promising navigation charts toward a better future.

The final moral certainty is that much will depend on politics. The devil is in the details, and any decisive steps will be preceded by complex and difficult negotiations between nations, regions, corporations, and many other groups. The world of 2100 will be shaped by millions of unpredictable decisions made in the next few decades. Though we are now a global species, it is not clear if our circles of concern have widened sufficiently to allow the sustained global collaboration needed to succeed as planetary managers. Whether or not we are able to do so will determine the fate of the new complex entity being born today on planet Earth: a conscious planet. The health and fate of that new entity will shape the futures of our descendants for centuries and will determine how long our lineage survives. Can we learn to collaborate as effectively as the cells that formed the first macrobes?

CHAPTER 9

Middle Futures

The Human Lineage

> *If we play our cards right, humanity is at an early stage*
> *of life: still in our adolescence; looking forward to a*
> *remarkable adulthood.*
>
> — TOBY ORD, 2020[1]

Distinctive Features of Middle Futures

Let's assume we survive the dangers of the next few centuries. What futures will our descendants face on scales of thousands and even millions of years? This is the "middle future." It is hard to think about middle futures with much rigor, because they will be shaped by unpredictable and purposeful beings like us. Besides, those futures stretch so far ahead that even quite regular trends will get lost in a fog of millions of other trends. But, as always, we have hints that are tantalizing and sometimes breathtaking, so I won't resist the temptation to explore what they may be whispering.

The middle future is less personal than the near future. True, we may feel some loyalty to the vast community of human-like creatures, as we imagine it stretching backward and forward in time. And it is fascinating to think that the future of that community could be much longer and richer than its past. But we can't have the same visceral concern for our remote descendants that we feel for those who will live through the next hundred years. Nor will we have as much influence over their lives. The consequences of today's actions will cascade chaotically down the centuries and millennia, like the flapping of a butterfly's wings, but we won't be able to say in any rigorous sense that they *caused* events in the remote future. Except in one crucial respect! If we don't survive the next few centuries, our descendants won't have any future at all. So the one thing we *can* do for our lineage is get through the bottleneck centuries during which we are learning how to manage a planet and have Armageddon weapons but few off-Earth settlements that can provide refuges in case of disaster. If we succeed, the generations living over the next few centuries will learn to live as part of a new complex entity, a conscious planet. That will open pathways to the middle future for billions of humans and post-humans. The chance of handing on such a spectacular legacy gives profound meaning to the times we live in now. If we get it right, the story continues.

The Next Thousand Years

Managing a Planet

Are conscious planets common? Are there millions in our galaxy? Or is what we are watching on Earth right now a cosmological anomaly? We don't know. But either way, the transition is immensely significant. It is one more example of the long cosmological trend of increasing complexity, the emergence of new structures built by rearranging existing things into new configurations with new properties. Is the crossing of this particular threshold the logical end point to the turbulent changes of the last few millennia of human history? Does it give a sort of meaning to human history? Could it be a step toward even larger complex entities organized at the scale of star systems or even galaxies?

What will it mean for humans to become planetary managers? We can imagine some plausible scenarios based on our first baby steps. Like all complex structures, a conscious planet will have distinctive emergent properties that will help it survive and flourish. It will certainly need: (1) coordination and planning at a planetary scale; (2) lots of good science and technology to solve wickedly complex problems; (3) new kinds of education so that most humans understand the collective challenges they face; and (4) ethical systems that motivate people to value a flourishing biosphere for the sake of their own descendants and billions of other species.

Coordinating and planning systems are already emerging, both within the United Nations and in the many international bodies, corporations, scholarly networks, and NGOs that manage international activities. Rates of scientific and technological progress are impressive today, and, except in the most pessimistic future scenarios, there is little reason to think they will slow. They might even accelerate. However, today's educational systems are not adapted to these new challenges, mainly because they are still locked in an era of nationalism and rarely teach a global vision. Education will have to be overhauled everywhere to provide the young with a planetary vision and the range of technical, political, and social skills needed to take part in the collective project of managing a planet. As H. G. Wells wrote, "Human history becomes more and more a race between education and catastrophe."[2] Educational systems will also be shaped by the evolution of planetary ethical systems. To succeed as planetary managers, coming generations will have to transcend tribal loyalties and learn to value not only humanity as a whole, but also the many other species with which we share this planet.

As humans start migrating to other planets and moons, we will experience much more hostile environments, and that will surely increase our respect for the beauty and hospitality of our home planet.

What long trends will shape the history of the next thousand years?

Political futures are almost impossible to imagine with confidence, though there exist some fascinating imaginary histories of the future.[3] We do know, though, that improved forms of planetary governance and coordination will have to evolve if we are to succeed as planetary managers.

Easier to imagine are technological trends, because the foundational trend of collective learning should sustain the technological creativity that has shaped human history so far. Besides, technologies have their own logic, and we can already see trends that may flourish in coming centuries. Important new technologies will include: (1) ways of producing lots of energy sustainably; (2) nanotechnologies; (3) artificial intelligence and robotics; and (4) biological technologies that will transform human bodies and turn some of our descendants into blends of humans and machines with indefinitely long lives.

New Energy Technologies

The modern world was built with the colossal energies of fossil fuels, but we now know we cannot keep using them. Can we preserve the gains of modernity by generating even more energy but in sustainable ways? We can be cautiously optimistic because many of the technologies we will need already exist. Two changes will be crucial: we must generate vast amounts of electricity sustainably, and we must use sustainably generated electricity to power everything from cars to manufacturing to communications to domestic appliances. The immediate challenge is to adopt those technologies fast, because in 2020 fossil fuels still accounted for about 85 percent of all energy used and shaped much of the world's infrastructure and most of its economic relationships.

Most promising energy-generating technologies are really new ways of mining sunlight. Hydropower taps solar energy indirectly by using flows of water created by evaporation and rainfall to drive turbines that generate electricity. Wind power uses air flows powered by the sun to drive turbines. Solar power mines sunlight directly. It collects solar energy via artificial forms of photosynthesis that are already more efficient than natural forms. The potential of these technologies is huge because their efficiency is improving quickly. We can imagine that within a century or two the world will be full of devices to capture the sun's energy — devices flexible and compact enough to be used on clothes, hats, rooftops, and roadways. Some will track the sun, like sunflowers. Hydrogen is also a promising source of energy, particularly for industries like aviation and steel manufacturing that require dense concentrations of energy. When hydrogen combines with oxygen it produces lots of energy, and its main waste product is water. The challenge is to find sustainable ways of producing and storing hydrogen.

In the twentieth century, there were high hopes for nonsolar energy generation from nuclear reactors. These faded after major accidents, such as the 1986 Chernobyl disaster, and because nuclear fission produces radioactive waste that can be toxic for millennia. But new, safer forms of nuclear power may still have a role to play.[4] Experiments with fusion power, which should be safer and cleaner than fission power, started in the mid-twentieth century. Here, the problem is to manage and contain reactions at extraordinarily high temperatures. But there are high hopes that these problems can be solved perhaps by the end of the twenty-first century. Of course, entirely new energy technologies may also emerge, such as the idea of launching vast numbers of satellites to collect solar energy in space and transmit it to Earth as microwave radiation.[5]

New technologies and new types of regulation may also make energy go further. Eco-taxes can discourage waste and shape how consumers use energy. Superconductors, which carry electricity with virtually no resistance, may drastically reduce losses of electricity in storage and transmission. Superconductors might also revolutionize transportation by powering supermagnets for virtually frictionless land travel.[6] At present, superconduction is only possible at very low temperatures, but there are hopes that room-temperature superconductors can be built within decades.

When refined and improved, like the James Watt steam engine, new energy technologies could lay the foundations for an era of abundant and sustainable energy production. That could be the first step to controlling much larger energy flows. In the 1960s, the Soviet astronomer Nikolai Kardashev suggested a novel way of thinking about energy control on cosmological scales. He was fascinated by the search for life in the universe (SETI, or the search for extraterrestrial intelligence) and wondered what sort of technologies were needed to transmit signals across galaxies.[7] To answer that question, he proposed a hypothetical ranking of civilizations based on the amounts of energy they mobilized. Type I civilizations control most of the energy reaching them from their star, or about 10^{17} watts. Modern humans consume almost that amount of energy, most of it from our sun, which is why Carl Sagan described the world energy system in the 1970s as equivalent to a Type 0.7 civilization on Kardashev's scale.[8] Within a century or two, new energy technologies will probably create a Type I civilization on Earth.

In Kardashev's hierarchy, Type II civilizations can control ten billion times as much energy, or about 10^{27} watts, by using most of the energy emitted by their local star. Doing that might be achieved by building "Dyson spheres," solar-system-size networks of solar panels that can mop up most of a star's energy output. That idea, first suggested by the novelist Olaf Stapledon in 1937 and explored more rigorously by the cosmologist Freeman Dyson in 1960, is fascinating, because if Dyson spheres already exist, it might be possible to detect them through the infrared light they would emit.[9] If we do detect them, they will represent entirely new types of complex entities, existing at the scale of whole star systems.

Type III civilizations on Kardashev's scale would represent complex entities at galactic scales. They can control another ten billion times more energy, or 10^{37} watts, which means being able to harvest energy on the scale of an entire galaxy. Some have speculated that the strange ring-shaped galaxies known as Hoag's objects might represent the galactic equivalent of a Dyson sphere. The idea is that such objects could have been built using a sort of galactic husbandry, by pruning or clearing away many star systems so as to create a gap between a galaxy's core and an outer ring of inhabited star systems. Civilizations in the outer ring could then use almost all the energy emitted from the core. Michio Kaku suggests that *Star Trek*'s Federation of Planets might represent a Type II civilization, while the Empire of the Star Wars series, which has colonized most of a galaxy, might be approaching the level of a Type III civilization.[10]

Kardashev's hypothetical ranking suggests ways of thinking about the technological levels our descendants might achieve and the new types of complex entities they might create. At present, though, there is little prospect of our descendants traveling to other star systems within the next millennium. So in the year 3000, however fancy their technologies, our descendants will probably remain at the technological level of a Type I civilization.

Nanotechnology: Tiny Machines

There are many promising developments in nanotechnology, or the building of machines at molecular scales. In a nanotechnology world, many of our machines may be as small as *E. coli* bacteria. The physicist Richard Feynman anticipated these technologies in a 1959 lecture called "There's Plenty of Room at the Bottom."[11] Since his time, nanotechnologies have blossomed. Computer chips are everywhere today, but now we can also move atoms around, one by one. The futurist John Smart argues that today most significant technological advances occur at the nanoscale, from the emergence of quantum computers to new materials such as carbon nanotubes, improved batteries, advances in superconducting, improvements in fusion technology, and genetic manipulation.[12] Biologists are already considering the possibilities of nanobots that can enter the body, cruise to problem areas, fix things, and then disassemble themselves like proteins inside a cell. Eventually, many machines will be invisibly small, and so cheap we will barely notice their cost in money or energy. Boosters of nanotechnology hope that within a few centuries our descendants will routinely build machines of fantastic power at negligible prices with no ecological impact.[13] By then, most manufacturing will no longer take place in factories, but in portable 3D nanoprinters that will be as ubiquitous as computers are today. Nanomachines will be all around us, and many will be inside us. Nanomanufacturing will play a particularly important role in off-Earth colonies.

Artificial Intelligence

Artificial intelligence (AI) and robots could prove even more revolutionary.[14] Early in the twenty-first century, we are already surrounded by machines, cars, weapons, and phones that are, by some measures, much smarter than we are, because they can calculate more precisely than us and analyze more information. In 2020, about half the world's population carried smartphones, each with far more computing power than Neil Armstrong's 1969 moon-landing craft, which had just sixty-four kilobytes of memory. Artificial intelligence has developed more slowly than many expected because it turned out that even very clever machines could not do some things that humans take for granted. They lack common sense because they take their logic too seriously, and they are not good at pattern recognition. These problems may be overcome in coming decades by teaching machines to learn by themselves through "deep learning." Computers have already taught themselves how to beat human world champions at games like chess and go. These machines don't aim at perfection, and they don't just calculate — like living organisms, they learn from experience, often by playing against themselves and recording the strategies that worked in the past. This is sophisticated future thinking. Often, their trainers do not understand the strategies they are using.

The big question for artificial intelligence research is whether we can keep control of smart machines once they are cleverer than us. A Spartacus rebellion by smart robots who enslave or exterminate our descendants is a terrifying prospect. It is also, in the estimate of Toby Ord (see Table 8.1, page 252), one of the more likely paths to an existential catastrophe. As early as 1863, Samuel Butler wrote, "We are ourselves creating our own successors. Man will become to the machine what the horse and the dog are to man."[15] Particularly scary is the thought that a robotic revolution could be launched and completed in nanoseconds. In *Superintelligence,* the philosopher Nick Bostrom imagines a network of computers dedicated to making paper clips and going about their task with remorseless single-mindedness. Eventually, they start converting the entire

Earth and then large parts of the observable universe into paper clips.[16] Can we prove that their goals are any better or worse than those of all living organisms (survive and reproduce)? We should also be worried by the fact that so much contemporary research into robotics and AI has been driven by military needs. Military robots are designed to kill. We can only hope that robotic warriors, missiles, and drones will be kept on a very tight leash for a very long time.

Scenarios like a violent robot rebellion are a reminder that human technologies could ruin or save us. But if we manage to keep clever machines under control, they will surely play a vital role in planetary management and the building of better futures. We are already forging powerful and pervasive alliances with computers and robots, in what is coming to be known as "the internet of things." Eventually, clever machines, including self-driving cars and prosthetic limbs, may be able to read their owners' thoughts. Implants that can respond to our thoughts are already available. In 1998, for the first time, an implant in the brain of a paralyzed man allowed him to control a computer by thinking, and similar implants can also be used to control wheelchairs and exoskeletons.[17] Will smart clothes powered by tiny embedded solar panels change texture or thickness in response to changes in the weather? A world full of intelligent and helpful machines, many too small to be visible, could seem commonplace within a few centuries. To many, their activities will seem as mysterious as the invisible magical forces that haunted the imagination of all ancient societies.

Transhumanism: Modifying Humans

New medical, biological, and genetic technologies will transform people as well as things. Indeed, if we were to visit the year 3000, many of the people we'd meet could seem as strange as the technologies they use and the cities they live in. In the late twentieth century, we learned how our genomes work, and we now know how to modify an organism's DNA gene by gene, using technologies such as CRISPR. Freeman Dyson has predicted that these powers will be used to manufacture new organisms and even nonbiological objects.[18] If we learn how to make artificial meat, that may transform the lives of the animals we currently treat as food and allow us to forge a more generous and compassionate relationship with many of our fellow species. If we learn how to grow houses, streetlamps, and vehicles biologically, that will transform the look of villages, towns, and cities. The blocky concrete cities of today will turn into Mandelbrot landscapes that look like vast gatherings of fungi.[19]

New biotechnologies will bring us close to Condorcet's dream of enhancing human bodies and extending human life spans. We are already modifying the genes of many other species, and it is only ethical reservations that have limited the routine modification of human genes. Already, embryos have been genetically edited. How long will it be before the birth of the first genetically engineered human baby, grown, perhaps in an artificial womb, and with an enhanced brain and no known genetic defects? By the year 3000 such procedures may seem as commonplace as using glasses or hearing aids today. Can we rule out the malevolent use of such technologies, perhaps to create whole underclasses to serve elites, such as the clones of Aldous Huxley's *Brave New World*?

The modern thought tradition known as transhumanism sees the enhancement of our own bodies as a welcome and perhaps dominant trend in human futures. Transhumanists look forward to cybernetic, biological, and genetic transformations that will enhance human physical and intellectual powers, spare people from most of the physical and psychological discomforts of life in human bodies, extend life spans indefinitely, and perhaps allow humans to blend seamlessly with machines. Natasha

Vita-More (her pseudonym is typical of the playful style of much transhumanist writing) lists some of the goals of transhumanism:

> At the core of transhumanism is the conviction that the lifespan be extended, aging reversed, and that death should be optional rather than compulsory. Transhumanism also proposes that artificial intelligence be used to help improve human level decision-making, that nanotechnology resolve environmental hazards, that molecular manufacturing stop poverty, and that genetic engineering mitigate diseases.[20]

In 1962, the writer and philosopher Stanislaw Lem imagined a world in which virtual-reality devices were so powerful, and connected with such precision to relevant parts of our brain, that we could no longer distinguish between real and virtual worlds. Thirty years later, he concluded that much of the technology already existed.[21] Will we eventually leave our human bodies entirely, perhaps by downloading our minds into computers, or into enhanced avatar bodies? Might we eventually move from body to body as, today, we move houses? Or share our consciousness with others directly? Can we imagine forms of education that make teachers redundant because knowledge is inserted into students' brains using future equivalents of today's USB drives?[22] Can we imagine legal systems that sentence criminals to brain modification?

Some of these ideas may make us squirm, but none can be ruled out. If the rapid medical developments of the last century or two continue, then it is likely that by the year 3000 many people will live healthy lives lasting hundreds of years and have powers that seem superhuman today. That will be the start of a slow division of our species into many subtypes. Within a few centuries there may be a growing diversity of humans, cyborgs, and transhumans, each with different special enhancements. For those of us living today, meeting such people would surely be an uncomfortable experience.

We can already see the early stages of most of the technologies discussed in this section, and we already live with some of them. But by the year 3000 it is likely that there will be many technologies we cannot even imagine today — technologies that would puzzle us as much as smartphones might have puzzled a resuscitated Cicero.

Migrating beyond Earth

Biological diversification will be accelerated by migrations to other planets and planetary bodies. Given the powerful ancient trend of global human migrations, we can be confident that many of our descendants will migrate into space in the next few centuries, taking other species with them. Once there, they will create new conscious planetary bodies in a sort of "budding-off" process — the conscious planet's equivalent of reproduction.

Ben Finney, a historian of the Pacific, writes:

> We evolved as an exploratory, migratory animal. . . . Our ancestors were able to spread from their tropical homeland through developing technology to travel to and survive in a multitude of environments for which they were not biologically adapted. Migrating into space, and the development of transport, life support and other systems which will be required for the spread and maintenance of human life there, represents a continuation of our terrestrial behavior, not a radical departure from it.[23]

Ben Finney has argued that Polynesian migrations into the Pacific offer helpful analogies for future human migrations into the solar system. Polynesian migrations depended on the existence of a vast archipelago of islands that provided stepping-stones into the Pacific like the many planets, moons, and aster-oids of our solar system.[24] And, like Polynesian migrations, future migrations through the solar system will depend on new navigational technologies, new types of ships, and the existence of groups willing and motivated to risk long and dangerous voyages and adapt to strange new environments.

Unlike the Polynesians, though, modern humans have sent robotic scouts ahead of us. Since the launch of the first human-made object to enter space (Sputnik, in 1957), we have sent hundreds of satellites to other parts of our system, and two, the 1977 Voyager satellites, have now left its outer edges. Within a century, thousands of humans may live on small colonies on the moon, on Mars, and even in mining colonies on asteroids. All will have armies of robots and 3-D printers to do much of the necessary heavy labor, building, and maintenance work. They will also bring other species with them, as Polynesians traveled with chickens, pigs, rats, taro, yams, and bananas. But the species taken to other planets may be genetically modified for different gravities and atmospheres. Building permanent settlements on the moon and Mars will be extraordinarily diffi-cult and costly, and many colonies will fail. But the task should not prove impossible. Conditions will be harsher than the harshest environments on Earth. But, unlike the first Polyne-sian migrants to new lands, we will arrive already knowing a lot from earlier robotic expeditions and with a glistening tool kit of new technologies.

Over several centuries, colonies will be established on many of the moons, planets, and asteroids of our solar system as well as on purpose-built starships, some the size of asteroids or small planets. A lot of manufacturing for humans still on Earth may migrate to space as well. After his first space flight in July 2021, Jeff Bezos, the CEO of Amazon, commented, "This sounds fantastical, what I'm about to tell you, but it will happen. We can move all heavy industry and all polluting industry off of Earth and operate it in space."[25] Like the earliest human migrants on Earth, those traveling to other parts of the solar system will try to change their new surroundings to make life easier. "Terraforming" will begin modestly, with protected shelters and underground warrens. But eventually colonists will try to transform landforms, build oceans, and create breathable atmospheres. Kim Stanley Robinson's Mars Trilogy imagines the terraforming of Mars, though he is probably optimistic in imagining large populations breathing Martian air within two or three centuries.

As they transform their new homes, interplanetary migrants will also transform themselves, culturally and biologically. Those who live permanently off Earth will think of Earth and humanity in new ways and will evolve new political systems, cultural norms, and technologies. They will change biologically, as they adapt to different atmospheres, pressures, and diets, to new circadian rhythms, and to the rigors of long journeys in space. They will also modify themselves using transhumanist technologies familiar on Earth.[26]

After the pioneering phase, planetary migrations will mark a significant new epoch in human history: the end of the brief but dangerous era that began in the twentieth century, when humans lived on a single planet but had the power to ruin their planetary home. That bottleneck era was when our lineage was most vulnerable. Migrations beyond Earth should increase the chance that our lineage will survive for hundreds of thousands, perhaps millions of years, just as biological reproduction allowed biological species to survive though individuals died.

Sometime late in the next millennium, after decades or centuries of robotic exploration, our descendants may head off to other star systems. They may use comets as stepping-stones, such as those in our solar system's Oort cloud, some of which are only loosely bound to our sun because they reach almost halfway to the nearest star, the triple star system of Alpha Centauri.[27] Our remote descendants may eventually colonize much of our galaxy just as the first living organisms on Earth once colonized Earth's young oceans. Interstellar migrations will depend on as yet unimagined technologies for driving ships, for maintaining sustainable environments, and for putting humans (or post-humans!) into hibernations lasting for centuries. Interstellar journeys will also depend on the existence of groups willing to risk long and dangerous voyages with little or no hope of returning. It would take spaceships traveling at 1 percent of the speed of light more than four hundred years to reach the Alpha Centauri system. But if they spread out from there at a similar rate, they could settle star systems throughout the Milky Way within one hundred million years, which is just a bit longer than the span of time since dinosaurs ruled our Earth.[28]

Will migrants from planet Earth encounter other life forms? The odds that they will seemed to improve in the late twentieth century, as we learned how common and diverse planets are in the universe, how vast are the interstellar clouds of life-forming molecules like amino acids, and how diverse are the conditions under which Earth-bound organisms can survive.[29] In a few decades we should learn enough about the atmospheres of nearby exoplanets to figure out if any contain life. But the chances of contacting complex, intelligent life-forms capable of some form of collective learning are much smaller. After all, it took more than three billion years for multicellular life-forms to flourish on planet Earth, and after sixty years of scanning the skies for messages, none have been detected so far. If we do come across other intelligent life-forms capable of collective learning, like us, the vast distances of space will probably keep us apart, as will biological, neurological, and technological differences. If we meet face-to-face, we are unlikely to meet species exactly at our technological level. The odds are that we will meet experienced travelers with technologies much more advanced than ours. We will be the greenhorns, and the technological gap will not favor us. This is a central idea in Cixin Liu's wonderful science fiction trilogy, *Remembrance of Earth's Past,* in which Earth is threatened by an invasion from the Alpha Centauri system. Of course, we might stumble on civilizations keen to protect species such as ourselves "as we preserve wild animals in national parks, for scientific interest."[30]

Scenarios

We can weave some of these threads together by imagining scenarios for the next thousand years as we did for the next century. A thousand years is a long time, long enough for different scenarios to play out many times in different eras and different regions of Earth and its colonies. But looking back one thousand years, or even two thousand years, to the time of Cicero, gives us some feeling for these time scales. So it may still be helpful to group possible scenarios under the labels we used in the previous chapter: "collapse," "downsizing," "sustainability," and "growth."

Extreme collapse scenarios mean the end of human history. That possibility will remain all too plausible during the next few "bottleneck" centuries, which Carl Sagan once described as the era of technological adolescence, in which we have the power to exterminate ourselves.[31] Indeed, over several centuries, the existential dangers may multiply. If we do go extinct, it will almost certainly be our fault. Our downfall could come from incompetence, lack of vision, inability to collaborate, or technological overreach and the creation of new powers we cannot control.

Less dire collapse scenarios include partial or regional collapses followed by a slow revival lasting many centuries. After a serious collapse — a fall-of-Rome scenario, for instance — recovery could take up much of the next millennium. Productivity will plummet, populations will fall to a few hundred million,

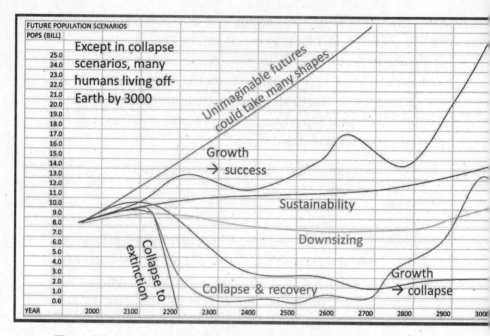

Figure 9.1: Populations over the Next One Thousand Years under Four Different Future Scenarios: Growth, Sustainability, Downsizing, and Collapse

and most people will live close to subsistence, though there may be protected pockets of affluence like feudal castles. Walter M. Miller's 1959 science fiction novel, *A Canticle for Leibowitz*, traces a centuries-long reemergence of new technologies after a nuclear war, remembered as the "flame deluge." Eventually (spoiler alert!), much modern technology is rediscovered, including nuclear weapons and . . . once again, the weapons are launched in a new flame deluge. Is it possible that we now live close to a sort of technological ceiling? Once Armageddon technologies exist, will they inevitably be deployed? That is an idea that would have depressed Condorcet and perhaps given Malthus a certain grim satisfaction. Such scenarios could explain why we have not had contact with intelligent aliens. It could be that we really are alone in the universe. But it could be that all intelligent life-forms capable of collective learning face a bottleneck era in which existential crises are almost inevitable, so they just don't last. Do species like us flicker in and out of existence in our universe like galactic fireflies?

Let's hope Miller's denouement is too pessimistic. We can also imagine scenarios in which humans, though slow learners, eventually learn how to manage themselves and planet Earth with some skill after apprenticeships lasting many centuries. Perhaps our descendants will live through generations of collapse, war, and revival before they acquire the political skills, the collaborative ethos, and the new technologies needed to manage planet Earth well. In that version of collapse scenarios, our descendants could just be graduating as competent planetary managers by the year 3000.

Other plausible scenarios are more optimistic. They envisage futures in which humans learn quickly how to manage a planet and get better at the job over time; in which reasonably effective global governance emerges; in which cybernetic, biological, and genetic modifications improve human health, extend life spans, and perhaps create new subspecies of humans; and in which technological innovation is sustained and human creativity generates many new ways of living and being. In all these scenarios, some humans will start living off-Earth, which will eventually bring the bottleneck era to an end.

In downsizing scenarios, most societies will be more ascetic and less consumer oriented than the wealthy societies of the early twenty-first century. Even in the year 3000, material living standards may not be much higher than those of today. But perhaps downsizing scenarios will only appear periodically in parts of the world, perhaps after major crises. Or perhaps they will play out mostly on off-Earth colonies where conditions are harsher.

"Growth" scenarios will persist with the competitive capitalist methods characteristic of the modern era. As we have seen, the main danger of such scenarios is that they underestimate the ecological dangers of unchecked growth in consumption and resource use. But if they work, they will generate dazzling new technologies, solve many of our current environmental challenges, and create societies of unprecedented material wealth. Even so, the capitalist engine of this type of growth also makes it likely that the world will become more inegalitarian, and that may create significant instabilities and conflicts both within and between different "nations" or regions. And it is likely that biodiversity will continue to decline in societies focused more on growth than on sustainability. Urras, the second of the twin planets in Ursula Le Guin's *The Dispossessed,* may offer a fictional model for a somewhat decadent growth scenario.

"Sustainability" scenarios will create societies close to those envisaged in the global Utopias of the early twenty-first century. Our descendants will rapidly come to terms with the long-term challenge of managing a planet and a biosphere. Global climates will be warmer than today for centuries, and levels of biodiversity will return very slowly, if at all, to those of the distant past, because even in the most optimistic scenarios humans will continue to consume a disproportionate share of Earth's resources. But rates of extinction of other species will stabilize. Greenhouse gas levels will be monitored closely, and experiences living in off-Earth colonies will remind our descendants constantly how important it is to respect ecological rules and resource limits.

In sustainability scenarios, Earth-bound populations a thousand years from now may be similar in size to those of today or even smaller, though total human populations will be larger because many humans will live in colonies off-Earth. Most people will live longer, healthier lives; many will live with biological enhancements; and many will be cloned or genetically enhanced. Differences in wealth will be limited, and most people on Earth, as well as those in the more benign off-Earth colonies, will have material living standards higher than ours are today, though not extravagantly so, as today's consumerist ethos will have long since been abandoned. Advanced forms of 3D printing will be ubiquitous, allowing people to print new organs, new gadgets, new houses, and new vehicles as easily as we print documents today. Many machines will be smart but tiny. Nanomachines will clean up waste, search our bodies for cancer cells, dispense vitamins and other medicines, cruise the oceans and atmosphere drawing down greenhouse gases, and hover in space as they help maintain spaceships and off-Earth colonies. All will have their own power systems. Many of the objects and buildings in towns, cities, and off-Earth colonies will be grown from organic materials, so there will be curves and soft textures where today we see straight lines and blocks.

In the most attractive sustainability scenarios there will be effective but nonauthoritarian global systems of governance, but much power will be devolved to regional or local levels. Growth in consumption will slow, but there will be spectacular growth in knowledge, science, and creativity. Art will flourish in forms and media we cannot imagine today. These will not be stagnant societies, and innovations of many different kinds will create diverse new lifeways, ways of thinking, and forms of sociability. The needs of off-Earth colonies will drive the development of new technologies for space transportation and communication, and the austere lifeways of off-Earth colonies may influence lifestyles, fashions, and ethical ideas on Earth.

The futures that eventually appear will combine elements of all these scenarios. The extremes — total collapse or Utopian success — are least likely. More likely are futures in which elements of different scenarios play out in different regions or in different periods. They will be shaped by unexpected contingencies, both good and bad, freakish disasters, sudden technological breakthroughs, and weird breakdowns, which will take us in directions we cannot imagine today.

The Human Lineage in the Distant Future

If we get through the bottleneck centuries, our descendants could be around for another few hundred thousand years (as long as we have existed so far) or even, perhaps, for millions of years. Speculating about futures at these scales takes us close to the realm science fiction aficionados call "space opera," in which, as Kim Stanley Robinson puts it, "you're zipping about the galaxy and the laws of physics are much relaxed."[32] Isaac Asimov set his Foundation series of novels fifty thousand years in the future, in an imagined realm of thousands of communities scattered across the galaxy. Given the vast distances between star systems, and assuming that we do not find out how to travel at light-speed or faster (which seems extremely unlikely at present), there will no longer be one collective future for humanity. Different future scenarios will play out around different stars, and there will be scenarios we cannot imagine today.

No longer will it even be clear what we mean by "humans," because transhumanist technologies and evolutionary adaptations to different environments will split humanity into many subspecies. That will end the brief moment in human history, which began about fifty thousand years ago, when there was only one species of humans. Our descendants will diverge biologically, technologically, and culturally as they settle diverse environments on planets, moons, and artificial satellites orbiting many different stars in different regions of the galaxy or cruising interstellar space. Kardashev's models suggest the possibility that the most technologically advanced societies will control much of the energy of stars, using technologies we cannot conceive of for purposes we cannot imagine.

The long trend of collective learning will surely continue. Our descendants will swap information and generate new technologies, new ways of controlling their environments and traveling, new ways of living together, and new forms of play, art, and spirituality. Networks of exchange will reach across star systems. That will guarantee some sharing of ideas, lifeways, and technologies, and also a fantastic diversity of human-descended civilizations. Our descendants may eventually travel to other galaxies. As they migrate, they may meet, trade with, fight, or even merge with other species also capable of collective learning. The current era of human history, when a single, homogenous species inhabits a single planet, will look like a momentary blip in a remote and fantastic past.

Trends like these provide the starting points for some of the most extreme speculations about remote post-human futures. Michio Kaku builds on Kardashev's typology of future civilizations to imagine Type III civilizations so powerful that they can bend what we think of as fundamental physical laws. Perhaps, he suggests, our remote descendants will reshape space and time by constructing "wormholes" to other universes if our own becomes too uncomfortable, too cold, or just too boring. In an influential 1979 essay, "Time without End," Freeman Dyson suggests that if life survives and continues to evolve as it has on Earth, "it is impossible to set any limit to the variety of physical forms that life may assume."[33] Life and consciousness could escape from flesh and blood and enter machines or assume even more exotic forms. One possibility Dyson imagined was life forms inhabiting deep space and built from structured clouds of charged dust particles, in a reference to the alien protagonist of Fred Hoyle's 1957 novel, *The Black Cloud.* Such evolutionary transformations might mean that life could survive in interstellar space for countless billions of years, even in a cooling and aging universe with little free energy and no liquid water. Perhaps, in such a universe, the living will live more slowly, and spend long periods in hibernation.

This of course is all highly speculative, and perhaps the best we can say of it is that it is not totally preposterous. But if they are even remotely plausible, such speculations remind us that the emergence of living beings capable of collective learning was momentous, wherever and whenever it has happened in our universe.

CHAPTER 10

Remote Futures

The Rest of Time

My battery is low and it's getting dark.
— JACOB MARGOLIS, JOURNALIST, FEBRUARY 2019;
A POETIC PARAPHRASE OF THE FINAL MESSAGE SENT FROM
MARS BY OPPORTUNITY ROVER ON JUNE 10, 2018[1]

Distinctive Features of Remote Futures

This chapter looks beyond the future of our lineage to the futures of the Earth, the sun, the galaxy, and the whole universe. It ends by speculating about the end of the epic story that began with the big bang 13.8 billion years ago. Now we seriously try to look through the eyes of the gods at the vast, sprawling, four-dimensional maps of B-series time and the block-universe. Our efforts are tentative and speculative, as always, but sometimes you get the spooky feeling that we can see the shape of the remote future more clearly than that of the next few centuries or millennia. And those blurry glimpses seem to be telling us something surprising and important: we live at the beginning of time. Our universe is young, and most of its story is still to be told.

We will have no personal connection to these remote futures, nor can we influence them in any significant way. But curiously, imagining the most remote futures can seem easier than imagining the middle futures of the last chapter. This is because the maps of remote time will be shaped mainly by pretty regular, mechanical processes, so we can spend less time worrying about the unpredictable behavior of purposeful beings like ourselves. Planets behave in an orderly way; so do galaxies; and so, it seems, does the universe as a whole, if you ignore the unpredictable activities of tiny numbers of purposeful beings. In this chapter our future thinking will work with large trends that seem mechanical, regular, and persistent enough for us to project them into the distant future with some confidence. Cosmologists tell us that there are even trends in the evolution of the whole universe that we can plausibly project to the end of time.

Of course, our confidence might be misplaced. All predictions could be rendered absurd by a discovery tomorrow, and what looks plausible today may seem laughable in twenty years.

Planetary and Galactic Futures

The Future of the Earth, Sun, and Solar System

Purposeful beings will have some role to play even on very large scales. For a few million years, the mix of species on our planet and its climate and ocean systems will be determined largely by the activities of humans. By the early twenty-first century, human activities had already reduced Earth's total biomass by as much as 50 percent, mostly through deforestation, and biodiversity is unlikely to recover fully as long as humans flourish on Earth.[2] If our lineage migrates to other planets and other star systems, our impact will be felt on galactic scales. And in some of the more spectacular scenarios of the last chapter, our remote descendants may acquire the power to manipulate entire galaxies and even to bend the laws of physics as we know them today. But in most scenarios for the remote future, our descendants will be bit players.

On scales of hundreds of millions of years, geological and astronomical processes will be the main shapers of the Earth's future.[3] The trends of plate tectonics are now well enough understood that we can make reasonably confident predictions about the Earth's geographical configuration one or two hundred million years from now.[4] Most tectonic plates move at a rate of several centimeters per year in known directions, so geologists can roughly predict how continents and oceans will rearrange themselves. The Atlantic is widening, but the Pacific and Mediterranean are shrinking, so it is likely that the Americas will meet up with East and Southeast Asia and Australia, while North Africa and Europe will find themselves soldered together over a vanished Mediterranean. In about two hundred million years, these trends will gather the scattered continental fragments of today's world into a new supercontinent that some have already christened "Amasia." It will be surrounded by a vastly expanded Atlantic Ocean. The last supercontinent, Pangea, began to fall apart about two hundred million years ago, so this regathering will be part of a long, cyclical trend that may have lasted for several billion years.[5]

In two or three billion years, the motion of tectonic plates will slow and stall as less heat is dissipated from the Earth's core. When the machinery of plate tectonics grinds to a halt, the Earth's geography will freeze into a planetwide tectonic rictus. Mountain building will cease, erosion will smooth continental surfaces, and winds will howl across vast plains with nothing to restrain them.

The long-term fate of Earth and the solar system will be governed by the evolution of our home star, the sun. That makes sense once you realize that the sun contains 99.8 percent of the matter in our solar system — the planets and moons orbiting it, including our own, are little more than a dusty halo. Traditional religions were surely right in thinking that the sun is the feudal lord of our small part of the universe. Other stars will have little impact on our future because the distances between stars are tremendous. If the sun were the size of a grapefruit (and planet Earth the size of a flower seed orbiting sixteen meters away), the nearest star system, the three-star system of Alpha Centauri, would be about forty-three hundred kilometers away (4.4 light-years), or slightly farther than San Francisco is from New York.[6] This sort of social distancing is typical for stars that lie, like our sun, about halfway between the Milky Way's dense core and its thinly populated outer borderlands.

The sun and solar system, like most stars in our galaxy, orbit the galaxy's central regions. Since it was born, 4.5 billion years ago, our solar system has circled the galaxy about twenty times, sometimes above and sometimes below its central plane, like a rider on a fairground merry-go-round, alongside millions of other stars. Neighboring stars will sometimes cozy up to us, but none will get too close. Once or twice every billion years, our solar system has passed through clouds of interstellar dust that blocked light from the rest of the universe for a few million years. If our remote descendants live through such a period, they will be tempted to think that our sun is the only star in the universe. At present, we are near the galaxy's central plane, so when we look toward the center of the Milky Way, our line of sight is blocked by clouds of interstellar dust. But in fifteen million years' time, we will be riding high above the plane, and any astronomers still on Earth will have a wonderful view of the Milky Way's central bulge.[7]

The habitual social distancing of stars means that we can imagine our solar system as a remote archipelago in an unimaginably vast ocean of stars and planets. Recently, we have met two strange travelers from other systems. The curious cookie-shaped object christened "Oumuamua" (Hawaiian for "a messenger from afar arriving first"), discovered in 2017, is probably a chip knocked off a Pluto-like planet from another star system.[8] Another interstellar traveler, 21/Borisov, was discovered in 2019 by an amateur astronomer in the Crimea. Both were traveling so fast they will escape the gravitational pull of our solar system, and that is why astronomers are confident that they are from other star systems.

Astronomers have studied millions of stars of different types at different stages in their evolution, so they understand the main trends of stellar evolution, and that means they can predict the future of our sun with some confidence.[9] Like all stars, our sun formed from a contracting cloud of hydrogen, helium, dust, and ice. It ignited when increasing pressure made the center of that cloud hot enough to fuse single protons into helium nuclei, and it will keep burning as long as it has a supply of unfused protons. Stars differ mainly in the size of the gas and dust cloud that birthed them. In larger clouds, gravity generates greater pressures and higher temperatures, which means that large stars burn through their stocks of protons fast and die sooner. Conversely, the small stars known as red dwarfs burn more slowly but will live for trillions of years.

Our sun is middling in size. It has lived for 4.5 billion years and will probably live as long again, unless, as some have suggested, our distant descendants, with Kardashev Type II or III technologies, learn how to domesticate it and stave off its death for a few billion more years.[10] Otherwise, our sun's long-term fate will depend on when it starts to run out of unfused protons. Over time, stars like our sun get larger and hotter as protons are fused to form helium nuclei that accumulate in their cores. Today, our sun is about 10 percent larger than when it was born, and it is emitting perhaps 40 percent more energy.[11] Life on Earth has survived this warming trend because, as solar emissions increased, the level of greenhouse gases in Earth's atmosphere fell, and that of non-greenhouse gases such as oxygen rose. Those changes helped maintain surface temperatures that allowed life to flourish because water could remain in liquid form in the oceans. Did we just get lucky or did living organisms help maintain the conditions they needed to survive on Earth, as James Lovelock has argued?[12] Whether he is right or wrong, nothing guarantees that the Earth will always be as life-friendly as it is today. Indeed, recent research suggests that in about a billion years, increasing heat from solar emissions could break down atmospheric carbon dioxide, which will be catastrophic for life because plants, which breathe carbon dioxide, will suffocate, and without the oxygen breathed out by plants, animals will suffocate too.[13]

In three or four billion years, an even hotter sun will boil away the oceans that are the original home of life on Earth. Radiation will split water molecules into hydrogen and oxygen atoms. Light hydrogen atoms will float off into space, while oxygen atoms will combine with materials such as iron, and the Earth will rust like an abandoned shipwreck. When surface temperatures reach one thousand degrees Celsius, rocks will start to melt like watches in a Salvador Dalí painting. Earth will start to look like its torrid and barren neighbor, Venus, as even the toughest of life-forms give up and perish.[14]

Within four billion years our sun will start showing its age. So far, it has been a stable and solid stellar citizen, and that has been good for life on Earth because it's meant that living organisms, including us humans, could depend on stable astronomical trends. Eventually, though, as the sun's proton stocks dwindle, it will start behaving more erratically. Fusion will slow and then, quite suddenly, the core will collapse. This is not yet the end because the sudden collapse will heat the core once more, and soaring temperatures will ignite fusion in regions outside the core. Those regions will expand until the sun's outer edges reach

Earth's current orbit. The sun will puff up and start shedding outer layers like discarded cardigans. By then, Earth and the sun's other planets may have retreated farther out as the sun's mass and gravitational pull will have declined. But Earth will surely get beaten up by the debris thrown off its increasingly unpredictable star. The gassy giant planets of the outer solar system, though at a safer distance, will also suffer from the sun's increasingly demented behavior. They, too, will drift away, and some may wander off into deep space to become homeless planetary nomads like our 2017 visitor, Oumuamua.

The sun will turn into a red giant. For several hundred million years it will look like Betelgeuse in the constellation of Orion. Now it will generate energy by fusing helium nuclei to form the nuclei of carbon, oxygen, and other elements. But forging large nuclei requires much higher temperatures than fusing single protons, so this phase may last just a few million years. In its final, mad death agonies, the sun will expand and contract in frenetic spasms.

Eventually, fusion will splutter and fail, and in about five billion years our shrunken, aging sun will die. Its corpse will glow for billions of years as a zombie star, a white dwarf. When it stops glowing, it will become a black dwarf, and it will hang around for a period much, much longer than it lived. If planet Earth has not been smashed up or incinerated by then, its frozen corpse may keep orbiting the dead star that once illuminated it. Or perhaps an unlucky slingshot may send it hurtling out of the solar system to wander for a virtual eternity between star systems along with billions of other wanderers like Oumuamua.

This may seem a bleak ending. But things could have been worse. Stars larger than our sun die young and die violently in supernova explosions so large that they could evaporate an entire solar system in hours. In such systems, there will not be time for life to get going. Our solar system is doomed to the much slower death of most average-size star systems, and something of its past will be preserved for countless billions of years as surviving planets dance in ghostly attendance around their dead sovereign. But in its distant past, this dead system will have once nurtured life, and not all star systems can say that.

Galactic Futures

Galaxies evolve, too, like stars and solar systems and people and bacteria.

Stars are born within galaxy-size creches made of hydrogen and a smattering of other elements. As stars die, they give back some of their raw materials when they explode in supernovas or throw off their outer layers. But the material in the cooling embers of dead stars is no longer available for star formation. This means that as galaxies age, they contain less and less material for making new stars. In our galaxy, rates of star formation may have peaked several billion years ago. The Milky Way is aging, and more and more of its inhabitants are star corpses, or red dwarfs. In a few billion years' time, reserves of star dust will dry up, and further star formation will depend on a tiny trickle of new material supplied by aging stars like our sun, as they cast off their outer layers. The era of energetic star formation that began a few hundred million years after the big bang will have ended, and galaxies throughout the universe will be dominated by small, slow-burning red dwarfs. Red-dwarf galaxies will last much longer than the dazzling galaxies of our era.[15]

Over time, gravity will draw neighboring galaxies into gentle but grandiose collisions. The Milky Way and Andromeda are the largest of the thirty to sixty galaxies in our local group.[16] Their gravitational pull will draw in smaller neighbors, stripping them of much of their stellar wealth. The Milky Way is already hauling in the two Magellanic Cloud galaxies and will slowly merge with them over hundreds of millions of years. As galaxies collide, individual stars will glide past one another, but their orbits will be deflected as gravitational fields warp and twist. Dust clouds will merge and become denser, which may trigger new bursts of star formation. Andromeda and the Milky Way are approaching each other at several hundred kilometers each second. In three or four billion years, as our sun approaches its final years, they will meet and either end up orbiting each other and spinning apart again, or slowly merging over millions of years. A merger will be slow but messy. The black holes at the centers of each galaxy will combine to form a new galactic monster. Stars will be sent spinning in new directions, which could carry them (and perhaps our dying sun) out into deep space or hurtling into the jaws of the monster at the merged galaxy's new core. Eventually, all the galaxies in our local group may form one colossal supergalaxy. Galaxy mergers will end as the accelerating expansion of the universe moves them too far apart to meet each other. Now each supergalaxy will seem like an isolated island in an increasingly lonely universe.

Cosmological Futures and the End of Time

How will it all end?

Surely all human societies have asked about the end of all things. There are two main possibilities: either the universe is eternal and has no end, or it is finite and does have an end. Many religious and philosophical systems of the Indian subcontinent imagine a universe without beginning or end.[17] So did modern scientific cosmology from Newton's time up to the middle of the twentieth century. But other traditions, including the Abrahamic religions of Judaism, Christianity, and Islam, imagine a universe created by God, with both a beginning and an end. And since the 1960s, most cosmologists have accepted the big bang paradigm, according to which, the universe had a beginning, followed by a long evolutionary history and an eventual ending. The name *big bang* was astronomer Fred Hoyle's scornful nickname for a paradigm he found absurd, and he continued to support the idea of a universe without beginning or end, a "steady-state universe." But most cosmologists now date the beginning of the universe to about 13.8 billion years ago, and many have speculated about when and how it will end.[18]

If you think there may be an end to time, you will imagine what theologians call an "eschatology," a grand, apocalyptic finale, perhaps, or possibly a slow fading away. "Eschatology is the study of the final end of things, the ultimate resolution of the creation's story."[19] According to some traditions, the end of time may be close, and some of us may even witness it. In such traditions, the last days will reveal the meaning of existence.[20]

Do scientific eschatologies have something to teach us about existence and the universe? Is there any scientific equivalent to the idea of "redemption"? There is, but only in the poetic sense that a glimpse of the final pages may tell us where we are in the story. Scientific cosmologies do not expect the end of time to reveal a purpose to the universe. But contemplating the end of time may tell us the overall shape of a story whose early stages we now know quite well. And that can be meaningful and aesthetically satisfying.

Modern Scientific Accounts of the End of Time

To grasp the scale of modern scientific eschatologies, we can take our sun's lifetime of about nine billion years as a basic unit.[21] So far, the universe has existed for just 1.4 solar lifetimes. Most cosmologists expect it to last for billions or trillions of solar lifetimes. That is meaningful. It tells us that we live in a young universe, near the beginning of time. So far, we have seen only the first few lines of the universe story. How will the rest of the story unfold? Are there cosmological trends that are regular enough and simple enough to give us some hints? Remarkably, there are, and despite the vast scales of modern cosmology, those trends mean that it may be easier to think about the end of time than to ask about the fate of humans in one hundred years' time.

All significant cosmological trends are linked to the idea that our universe is expanding. That idea was first proposed in the 1920s. The astronomer Edwin Hubble showed that distant galaxies were moving away faster than those closer to us, and, as the cosmologist Georges Lemaître pointed out, that must mean that the universe is expanding. For several decades the idea of an expanding universe lurked at the fringes of scientific cosmology. Slowly, though, evidence accumulated showing that the universe really had been different in the distant past, so it must have a history of some kind, and that suggests it must have had a beginning. In the late 1940s, the Russian American physicist George Gamow argued that if indeed there had been a big bang, then, as the universe expanded and cooled, there should have come a turning point when it cooled enough for charged protons and electrons to link up and form electrically neutral atoms. Suddenly, most of the matter in the universe would have become electrically neutral, and light energy would have been released in a huge flash. Few took the idea seriously until two radio astronomers, Arno Penzias and Robert Wilson, accidentally found Gamow's flash of energy in 1964. Today we know it as the "cosmic background radiation." That discovery clinched the idea of an expanding universe for most astronomers, and in the 1960s, "big bang cosmology" became the central paradigm of modern astronomy.

Big bang cosmology revolutionized astronomical thinking. It meant that the universe, like planet Earth and your neighbor's dog, had a history of change, and that encouraged a search for large trends in the universe's history. Would the expansion continue forever or would it eventually slow and perhaps even reverse? Answering that question required two types of measurements: measurements of how fast the universe was expanding and estimates of how much stuff there was in the universe. That second measurement was necessary because knowing how much mass the universe contains tells you how great a gravitational pull that mass exerts on the expansion. Is it enough to rein it in?[22] Is the trend of expansion perhaps slowing? If not, the universe will expand forever. It will cool down increasingly slowly, and its matter and energy will be scattered over vaster and vaster spaces and take increasingly disorganized forms. Entropy, having enabled the flows of free energy that allowed the building of complex structures, will eventually ensure their breakdown, so the universe will get simpler and simpler. That is the scenario that nineteenth-century physicists described as the "heat death" of the universe. Eventually, all energy will assume the random, structureless form of heat and the universe will run out of creativity. At best, a few scattered things and energies will jiggle randomly and meaninglessly forever and ever.

On the other hand, if there is enough mass to rein in the expansion, perhaps the universe will one day start contracting again. As it does, it will become denser and denser and hotter and hotter until all its matter and energies are once more concentrated into a tiny space similar to the original big bang. Will the whole cycle start again at that point, with a new big bang and a new universe? Could successive universes oscillate between two states, one cold, vast, and empty, the other hot, tiny, and packed with matter and energy? The idea of a collapsing universe also suggests other intriguing possibilities. If time travels in the direction of expansion, would time reverse direction in a collapsing universe? Stephen Hawking toyed with that idea in his bestseller, *A Brief History of Time*, but later dropped it.[23]

Several decades' worth of research failed to resolve these issues because astronomers found that the universe seemed poised right at the borderline between continuing expansion and eventual collapse. There seemed to be enough mass in the universe to eventually halt the expansion, but not quite enough to reverse it. That suggested that the universe would continue expanding forever, but more and more slowly. And that raised the deep, almost theological, question of why the universe should be so exquisitely poised at this strange cosmological point of balance, like a pin that won't fall over.

An unexpected resolution to these debates came in 1998, when two teams of astronomers led by Brian Schmidt in Australia and Saul Perlmutter in the United States tried to measure the universe's rate of expansion more accurately. They did so by studying type 1a supernovas, which all emit about the same amount of energy, so variations in their apparent brightness as seen from Earth can be used to measure their real distance with great precision. Both teams reached the astonishing conclusion that the universe's rate of expansion was not slowing; it was increasing and had been increasing for several billion years. We still do not understand why this should be so, though most cosmologists argue that there must be a form of energy (known as "dark energy" because no one understands it) that increases the rate of expansion as the universe gets larger. These findings, which are now accepted by most astronomers, suggest that the universe will expand faster and faster forever and ever.

If these ideas are right, the universe will get larger and larger, colder and colder, and emptier and emptier faster and faster, until different parts of the universe start to lose touch with one another because light from remote objects can no longer reach them. Eventually, we will no longer see anything beyond our own local group of galaxies, which would be held together by gravity. Any surviving astronomers may be puzzled by ancient records claiming that the universe contained not just a score of galaxies, but billions of them. In a fragmented universe ten thousand times as old as ours and much, much larger, even red dwarfs will die. Galaxies will consist then of dead stars, black holes, and the sort of random stuff that can pop out of the vacuum according to quantum physics. Black holes will gobble up the remaining star matter and there will come a time when bloated black holes and odd bits of quantum debris are more or less all that remains. Eventually, even black holes will evaporate, and all that will be left will be empty space and dark energy, perhaps a few forlorn photons of energy, and the odd stray particle wondering why it's there.[24] And then we (metaphorically, of course, because none of us will be there) will have to imagine inconceivably vast eons, in which a few photons and neutrinos wander forlornly without ever meeting anything else in a universe that is getting emptier and emptier, but continues a sort of zombie existence for a time beyond imagining. Nothing will last. We cannot even be sure that space and time will exist in any sense that we can understand. And yet, quantum physicists assure us, on virtually infinite time scales everything that could turn up may still pop briefly out of the void, including, perhaps a copy of your brain or a vase of tulips.[25] This is territory so weird that we should no longer take our speculations too seriously!

How likely is this story? It is only two decades old, and, as Jim Holt puts it, "cosmologists tend to reverse themselves every decade or so."[26] On today's evidence, this is the best story we can tell about the remote future of the universe. But there are many reasons for thinking the story will evolve. One is our ignorance about most of what is in the universe. The motion of most galaxies suggests that there must be a vast amount of matter that we cannot yet detect. Astronomers refer to it as "dark matter," and it may account for as much as 25 percent of the mass of the universe. Even more puzzling is the "dark energy" that seems to be accelerating the universe's rate of expansion. There are theoretical hints in Einstein's general theory of relativity about such a force. If it exists, then by now it probably accounts for about 70 percent of the mass of the universe. Add the mass of dark matter and dark energy and it looks as if we have no real understanding of 95 percent of the mass of our universe. Clearly, we are missing some important things. A better understanding of dark energy and dark matter in coming decades could change all bets about the future of the universe.

Another idea that could rewrite the story I have told is that of the "multiverse." Today, many cosmologists take seriously the possibility that our universe is not alone. There are good theoretical reasons (but no hard evidence yet) for imagining that big bangs may be occurring all the time in a huge multidimensional space even larger than the block-universe. This is what cosmologists call the multiverse. If so, it is possible that different universes, like different species of animals, have slightly different fundamental attributes, which might mean that different species of universes could evolve over some sort of pan-dimensional cosmological time. Perhaps gravity is slightly stronger, or electromagnetism slightly weaker in other universes.[27] In this scenario, there could be very different types of universes. Some may last just seconds. Some may last much longer than ours. Some may allow the creation of complex entities such as *E. coli* or rabbits, while others may not even allow the creation of stars. If, as physicist Lee Smolin argues, new universes might be created within black holes, that would mean that only those universes that can generate stars large enough to collapse into black holes would be able to reproduce and pass on their cosmological parameters to new generations of universes.[28] Over time, that seems to mean that such universes should become increasingly common through a sort of cosmological natural selection. And of course, only those universes capable of creating complex entities such as ourselves could be inhabited by creatures like ourselves, so there may be nothing strange about the fact that we find ourselves in a universe so finely tuned to allow complex things to emerge.

Lovely ideas, but . . . at present we have no evidence for

them. Our technologies allow us to observe only one universe. Any theories about other universes are based on logic and imagination alone. We have a sample of one. Astronomers used to set their students the following exam question: "Define the universe and give two examples."[29] For those of us trying to understand the future, the joke is painful. Though we constantly imagine a sort of "multifuture," in fact we only ever meet one.

Acknowledgments

Much of this book was written in Sydney during the COVID-19 pandemic, which has turned us all inward. In my case that has meant turning inward toward a loving family. Chardi and Emily put up with me vanishing into my study for long periods when there were probably better or more interesting things we could have been doing. Emily's scrumptious daughter, Sophia, has lit up our lives since she arrived (five months ago as I write this). Modern technologies have let us stay in regular contact with our English and American families, too, including Sophia's cousins, Daniel and Evie Rose, and their parents, Joshua and Olivia. And with our brothers and sisters, Diana, Rob, Russ, Fred, and Joe, in England and the United States. And with cousins and friends in England, the United States, and elsewhere. The slow gestation of this book was warmed by the feeling of being embedded in kind networks, while the abstract idea of the future was made real by thinking about *their* futures.

So many of my debts are to other scholars. I want to thank Macquarie University in Sydney, where I spent most of my academic career, for supporting the strange but powerful idea of big history for many years. The History Department at San Diego State University also supported the teaching of big history, and I have good friends in both departments. I must thank my big-history students over the years because, to an extent that few of them can imagine, it was conversations with them that really helped me understand the shape and power of the big-history story. Members of the International Big History Association have built a diverse and supportive community for those of us interested in scholarship that is as wide as can be, and Bill Gates and colleagues have generously supported the teaching of big history through the Big History Project.

The Brockman agency has given generous, enthusiastic, and always efficient support to the publication of two linked books, first *Origin Story* (on the past), and now *Future Stories* (yes, on the future!). Tracy Behar and Ian Straus at Little, Brown did two very detailed rounds of editing, which did a huge amount to spruce up my original manuscript. I am immensely grateful to them.

Writing a book like this takes you way beyond any formal scholarly expertise you may once have had. (Mine was in Russian history, which does leave one or two traces in this book.) That makes the comments of scholarly friends with expertise in other fields peculiarly important. In field after field I have asked the scholarly advice of friends about what I should be reading (and *not* reading because there is just not enough time to read the wrong books if your questions cross many disciplines). Those I have asked have been generous with their time, their comments, and their expertise. Many ideas and references have been picked up in brief conversations or email exchanges. And for those I offer my general, but warm thanks.

I particularly want to thank those who read part of this book as it evolved. The Australian futurist Joe Voros was my guide and mentor as I entered the strange but fascinating world of futures studies. He led me to the foundational works in the field. But his own work, which links big history and futures studies, was also an inspiration to me. And Joe offered very helpful comments on the manuscript at a very late stage. I also sent drafts to other friends and scholars, and all responded (often despite heavy workloads) with generous comments, often saving me from egregious blunders of fact or tone or emphasis. They include the astrophysicist Charlie Lineweaver; world historians Merry Wiesner-Hanks, Marnie Hughes-Warrington, Craig Benjamin, and Esther Quaedackers; a biologist, Michael Gillings; a philosopher and political scientist, Sasa Pavkovic; and my PhD student, Max Barnett. Charlie reminded me that entropy is a creative as well as a destructive force; Merry picked up surviving traces of Eurocentrism in my writing (ouch!); Craig picked up inconsistencies in terminology and in transliterations from Chinese; Michael cautioned me about attributing purpose too confidently to living organisms; and Sasa poured cold water on the more optimistic future scenarios in chapter 8. I did not follow all their suggestions, so it is particularly important to stress that I alone am to blame for remaining blunders, errors, infelicities, blind spots — all the things an author dreads discovering just after delivering the final manuscript. Saying this is particularly important when discussing a subject as strange as the future, because I often succumbed to the temptation to explore odd byways, sometimes against kind advice. My friends couldn't always save me from my own pigheadedness.

Glossary

Note: The glossary contains two main types of entry: (1) simple explanations of technical terms such as *entropy;* (2) definitions of terms used idiosyncratically in this book, such as *collective learning* or *future thinking.*

action potential: Electric pulses sent by neurons through axons to communicate with other neurons or with muscles; action potentials are powered by **chemiosmosis.**

agrarian era of human history: From the end of the last ice age, roughly ten thousand years ago, to the beginning of the modern era, a few centuries ago.

Anthropocene epoch: The era, beginning in the twentieth century, when our species suddenly became one of the major drivers of change on planet Earth. The idea was first proposed by the climate scientist Paul Crutzen in 2000 and has been taken up by scholars in many different fields and disciplines.

arrow of time: The idea that time moves in only one direction.

A-series time: A phrase that originated in a famous article on time by J. Ellis McTaggart to describe one of two broad approaches in the philosophy of time. A-series time takes seriously our everyday experience of time as a dynamic flow in which the future turns into the present and then into the past. The core metaphor is of time as a river. See **B-series time.**

big history: Histories of the whole of the past at multiple scales and through multiple disciplinary lenses.

block-universe: An imagined entity containing all of the past and future of the entire universe; implicit in **B-series time** and sometimes thought of as a god's-eye view of time.

B-series time: A phrase that originated in a famous article on time by J. Ellis McTaggart, to describe one of two broad approaches in the philosophy of time. B-series time transcends our everyday experience and views time, as it were, from above. The core metaphor is of time as a map. See **A-series time; block-universe.**

causation: The idea that one event may explain other, later events. The idea is essential to most forms of future thinking because it allows us to predict consequences if we observe causes. But the idea of causation leads to difficult philosophical conundrums because, as Hume showed, it is impossible to prove conclusively that A caused B, as every B has multiple causes. Recent work (e.g., by Judea Pearl) shows that the idea of causation remains indispensable and powerful if causation is understood as the result of local, perspectival, and probabilistic interventions in events.

cell: The basic building block of all living organisms; a semiporous membrane surrounds all the molecules and substances needed for a cell to survive. See **eukaryotes; prokaryotes.**

certainty: As defined in this book, comes in two forms. *Absolute certainty* allows of no exceptions and is probably unattainable in the real world, except within deductive chains of reasoning. *Moral certainty* describes assertions, including assertions about the future, that we trust sufficiently to act on. See **moral certainty.**

chemiosmosis: Most (possibly all) cells can pump positive ions outside their membranes, creating a small negative charge that can be used to power important biochemical processes, including the sending of electric pulses to other cells. See **action potential.**

circadian rhythms: Internally generated rhythms that help organisms keep track of rhythms in the external world. They are probably present in all cells and organisms.

circumnutation: A term used by Darwin to describe the random circling motions that plants use to explore their environments.

clock time: The dominant modern sense of time as universal, metronomic, measurable. Because we often have to align our own rhythms to clock time, it can feel coercive.

collective learning: The ability, unique to humans and enabled by human language, to share, store, and accumulate new ideas, experiences, insights, and information with such precision and on such a scale that the information available to humans increases on historical time scales. Collective learning explains why human control over many aspects of our environments and futures has increased in the course of human history, until today we dominate change on planet Earth. See **Anthropocene epoch.**

complex entities: Structures composed of diverse components arranged in precise ways that give them distinctive "emergent" properties, and so organized that they can maintain their integrity for a period of time.

deduction: Logical arguments that lead to certain conclusions if based on axioms known to be true; most evident in mathematical reasoning. See **induction.**

determinism: In extreme forms, determinism tells us that every detail of the universe's history was in principle predictable from the moment of the universe's creation.

disenchantment: The idea of the sociologist Max Weber that a key feature of modern science is its rejection of the presence of arbitrary or capricious spirits, forces, and gods, leaving a universe shaped mainly by regular, mechanical processes that allow some degree of prediction.

divination: Attempts to perceive or change aspects of the future through contact with spirit beings or gods or forces.

energy: The forces that have the potential to make things happen. See **free energy; second law of thermodynamics.**

entropy: A measure of disorder; see **second law of thermodynamics.**

eukaryotes: Cells that include a nucleus and other organelles; all macrobes are formed from eukaryotic cells. See **prokaryotes.**

forecast/forecasting: Any attempt to identify likely futures. Used here as a synonym for **predict/prediction.**

foundational era of human history: The first and by far the longest era of human history, from the evolution of the first humans, several hundred thousand years ago, until the end of the last ice age, roughly ten thousand years ago. This era is often described using other labels such as "Paleolithic"; the label used here stresses that this is when the foundations were laid for all later eras of human history.

free energy: Energy, like gravity, that flows in orderly, nonrandom ways so it can do work and make things happen or change.

future: All of time except for the past and present. In A-series time, the future includes many possible futures. We do not know in what sense the future "exists" in the present, so strictly, the future refers to our present *anticipations* of what may lie in the future.

future cone: A type of diagram of possible futures based on diagrams first developed by Einstein and Minkowski.

future management: Using future thinking to intervene in events in the hope of steering toward our preferred imagined futures or **Utopias.**

futures studies: The broad, transdisciplinary field of study of possible futures that emerged in the twentieth century.

future thinking: Used broadly in this book to refer to all the methods living organisms use to prepare for and cope with uncertain futures. The phrase covers many different terms used by futurists, such as *foresight, forecasting, prediction,* and *prognostics.*

induction: Logical arguments that draw conclusions about what is unknown from what is known. Unlike deductive arguments, inductive arguments always involve a leap of faith. Most forms of **future thinking** are based on induction. See **deduction.**

information: Information about the past and present is crucial for future thinking because it reduces uncertainty and limits the number of possible futures. As a general rule, more (good) information increases our power to predict.

living organisms: Life consists of living organisms, complex entities made of cells, that seem to show a distinctive purposefulness and creativity as they try to maintain their structures and reproduce despite the destructive forces all around them.

LUCA: Last universal common ancestor; an imagined ancestor of all life on Earth, that probably existed almost four billion years ago.

macrobes: Organisms, such as ourselves, that are built from large numbers of eukaryotic cells. Also known as **multicellular organisms.**

matter: The physical "stuff" of the universe that occupies space. Einstein showed that matter consists of compressed energy and can be converted back to energy, for example, by proton fusion. See **energy.**

mechanical behavior: Processes of change of all kinds that can be explained as the result of mechanical laws acting on passive, purposeless entities, so they are usually regular enough to allow some degree of forecasting. See **purposefulness.**

microbes: Single-celled organisms including all prokaryotes.

middle futures: Imagined futures at scales of hundreds, thousands, even millions of years.

modern era of human history: Recent centuries, in which globalization, new technologies, and energy from fossil fuels have given rise to today's global, planet-changing human society.

moral certainty: Sufficient certainty about a probabilistic process, including processes in the future, to allow action. The phrase was used by Leibniz.

multicellular organisms: Used here as a synonym for **macrobes.**

Nasreddin Hoca strategy: The paradoxical research strategy of trying to learn about something for which we have little or no direct evidence by looking somewhere else where evidence exists that may be relevant. Based on a story about a famous Turkish sage.

natural time: The rhythms of the nonhuman world, such as night and day or seasonal changes.

near future: Imagined futures at scales of about one hundred years.

predict/prediction: Often used (critically) of overprecise or overconfident assertions about likely futures but here used as a synonym for **forecast/ forecasting.**

probability: The likelihood of an event occurring or a claim proving to be true; probability theory, from the seventeenth century, put the calculation of probabilities on a rigorous mathematical footing that is, today, the basis of all statistical thought.

prokaryotes: Cells that lack a nucleus and organelles; most single-celled organisms are prokaryotic. See **eukaryotes.**

protein: Molecules present in all cells. They consist of precisely ordered chains of smaller molecules (amino acids) that fold into very precise structures that enable proteins to do much of the fundamental biochemical work of cells.

psychological time: The erratic temporal rhythms of our bodies and minds, such as those of alertness and tiredness, wakefulness and sleep.

punctuated equilibria: Patterns of change first identified in biological evolution but present in the evolution of all complex entities. An emergence phase is followed by a relatively stable phase and eventually by a phase of collapse and breakdown.

purposefulness: Here used to describe behaviors that *look* as if they are driven by purposes such as survival and reproduction. All living organisms exhibit such behavior, but we do not really understand its source. Purposeful behavior, including that of humans, usually seems too irregular to allow confident forecasts. See **mechanical behavior.**

random dipping: a technique for imagining possible futures based on random choices in the present, such as the throwing of dice.

remote futures: Imagined futures at scales of billions of years, up to the lifetime of the universe.

second law of thermodynamics: Not strictly a "law" because it is probabilistic; it asserts that in a closed system (such as the universe), entropy (disorder) tends to increase, which suggests the existence of a direction to time, at least for complex entities whose structures will break down some time in the future. Paradoxically, the second law explains both the flows of energy that allow the emergence of complex entities, and the fact that all complex entities will eventually break down. See **entropy.**

social time: The rhythms imposed on our activities by those of other humans. Social time has become increasingly significant as human networks of exchange have increased in size and significance.

sweet spot: The delicate point of balance, aimed at by all serious forecasters, between excessive generality (which makes forecasts vacuous and uninteresting) and excessive precision (which ensures they will prove inaccurate).

trend hunting: Attempts to identify and understand regular trends in the past that can be used as clues to likely futures.

Utopia: Used in this book to refer to the goals that organisms aim for as they try to navigate the future.

Zone of Anxiety: A domain of imagined futures in which we care deeply about what may happen and suspect that forecasting may be possible; this is the zone of possible futures that generates most efforts at forecasting and prediction.

Notes

Introduction

1. Corinthians 13:12 (Authorized [King James] Version).
2. Cicero, *On Divination*, bk. 2, 395.
3. Collingwood, *The Idea of History*, 120.
4. Rescher, *Predicting the Future*, 1.
5. There are, however, some textbooks on modern approaches to planning for the future, such as Hines and Bishop, *Thinking about the Future*, and Szostak, *Making Sense of the Future*.
6. On big history, see Benjamin, Quaedackers, and Baker, *Routledge Companion to Big History;* Christian, *Origin Story;* Gibelyou and Northrop, *Big Ideas;* on big history and future thinking, see Voros, "Big Futures."
7. Garrett, *Hume*, xix.
8. Watts, "'New' Science of Networks," 243–46, and Caldarelli and Catanzaro, *Networks*, 2; on the Silk Roads, see Christian, "Silk Roads or Steppe Roads."
9. Bell, *Foundations of Futures Studies*, 1:182.
10. Schrödinger, *What Is Life?* 1.
11. Collingwood, *The Idea of History*, 54; Carr, *What Is History?* 68–69; Confucius cited in De Vito and Della Sala, "Predicting the Future," 1019.
12. See Christian, *Maps of Time* and *Origin Story*.

Chapter 1: What Is the Future?

1. Cited from Guthrie, *Faces in the Clouds*, loc. 1056, Kindle.
2. Dator, *A Noticer in Time*, 77.
3. "Temporalities," forum in *Past and Present;* Wood, "Big History and the Study of Time."
4. From Holt, *When Einstein Walked with Gödel*, 20.
5. Omar Khayyám, *Rubaiyat*, xxvii.
6. Milton, *Paradise Lost*, bk. 1.
7. Augustine, *Confessions*, bk. 11, 230, 238; Ismael, *How Physics Makes Us Free*, 210.

8. Ismael, "Temporal Experience," 460; good introductions include Bardon, *Brief History of the Philosophy of Time*, and Baron and Miller, *Introduction to the Philosophy of Time;* and see Callender, *Oxford Handbook of Philosophy of Time*.

9. T. R. V. Murti, cited in Loy, "Māhāyana Deconstruction of Time," 14.

10. Mellor, *Real Time* and *Real Time II*.

11. Baron and Miller, *Introduction to the Philosophy of Time;* McTaggart, "Unreality of Time" (1908).

12. Bardon, *Brief History of the Philosophy of Time*, 6; McTaggart, "Unreality of Time" (1908), 458.

13. Newton wrote in Latin; I have cited the first English translation, by Andrew Motte, published in 1729.

14. Mark Twain, *Adventures of Huckleberry Finn* (New York: Charles L. Webster, 1884), chap. 12, https://etc.usf.edu/lit2go/21/the-adventures-of-huckleberry-finn/141/chapter-12/.

15. Omar Khayyám, *Rubaiyat*, xxvi.

16. Cossins, "The Time Delusion," 34; Dator, *A Noticer in Time*, 79, on Hawaiian time; McGrath, "Deep Histories in Time, or Crossing the Great Divide?" 4.

17. The Australian futurist Joe Voros introduced me to future cones; see Voros, "Big History and Anticipation."

18. Price, *Time's Arrow and Archimedes' Point*, chap. 1.

19. James, "The Dilemma of Determinism," in *Delphi Complete Works of William James*, loc. 36,352, Kindle.

20. Augustine, *Confessions*, bk. 11, 228; Blackburn, *The Big Questions: Philosophy*, loc. 1720, Kindle.

21. Isaacson, *Einstein: His Life and Universe*, loc. 9621, Kindle.

22. Cited in Gallois, "Zen History," 432–33. My thanks to Ian Straus for reminding me of the Tralfamadorians.

23. Augustine, *Confessions*, bk. 11, 232.

24. James, "Perception of Time," chap. 15 of *The Principles of Psychology*, in *Delphi Complete Works of William James*, loc. 11,732, Kindle; Dennett, *Consciousness Explained*, chap. 5 on relevant psychological experiments, e.g., 111.

25. Augustine, *Confessions*, bk. 11, 233.

26. Rynasiewicz, "Newton's Views on Space, Time, and Motion," 1; Newton quote is from Westfall, *Life of Isaac Newton*, 259. Newton later tried to retract the metaphor of a "sensorium" but always insisted that God was "omnipresent in the literal sense."

27. Omar Khayyám, *Rubaiyat*, trans. Edward Fitzgerald, 5th ed., 1889, no. 73.

28. Laplace, "Philosophical Essay on Probabilities," 3–4.

29. Cicero, *On Divination*, bk. 1, 361–63.

30. Augustine, *City of God*, bk. 5, chap. 9, vol. 412, 177, in a section arguing against Cicero.

31. For Laudan, see Curd and Cover, *Philosophy of Science*, 152; and see Hacking, *The Taming of Chance*.

32. On Russell, see Paul Davies, *Demon in the Machine*, 68ff.; Waldrop, *Complexity*, 328.

33. Gisin, "Mathematical Languages Shape Our Understanding of Time in Physics"; his arguments are summarized in Wolchover, "Does Time Really Flow?"

34. Feynman, *Character of Physical Law,* lecture 6, loc. 1981, Kindle.

35. Boethius, *Consolation of Philosophy,* 159.

36. Anderson, "More Is Different"; William James, "The Dilemma of Determinism," in *Delphi Complete Works of William James,* loc. 35,914, Kindle.

37. Waldrop, *Complexity,* loc. 774, Kindle.

38. Hume, *Treatise of Human Nature,* pt. 3, sec. 2; Baron and Miller, *Introduction to the Philosophy of Time,* chap. 6; Russell, *History of Western Philosophy,* 85; on smoking and lung cancer, McGrayne, *Theory That Would Not Die,* 112.

39. See Kistler, "Causation," who cites Russell, "On the Notion of Cause."

40. Russell, "Psychological and Physical Causal Laws," 288–89.

41. Pearl and Mackenzie, *The Book of Why;* Pearl, "Art and Science of Cause and Effect."

42. I used a reversed film of an egg being scrambled in a 2011 TED Talk, "History of the World in 18 Minutes," filmed March 2011 in Long Beach, CA, TED video, 17:24, https://www.ted.com/talks/david_christian_the_history_of_our_world_in_18_minutes?language=en.

43. On the paradoxical relationship between complexity and the second law, see Egan and Lineweaver, "Life, Gravity and the Second Law of Thermodynamics."

44. Price, *Time's Arrow and Archimedes' Point,* chap. 3.

Chapter 2: Practical Future Thinking

1. Ismael, "Temporal Experience," 480.

2. Marx, *The Marx-Engels Reader,* 145.

3. Quotations from Bhagavad Gita, discourses 1, 2, 3 and 11.

4. Einstein, "On the Electrodynamics of Moving Bodies."

5. Elias, *Time,* 4.

6. Strictly, that's the speed of light in a vacuum.

7. Quotation cited from Riggs, "Contemporary Concepts," 51; Einstein, *Relativity,* chap. 9.

8. Danks, "Safe-and-Substantive Perspectivism," 127.

9. Wilczek, *Fundamentals,* 188ff.

10. E.g., Christian, *Origin Story.*

11. Nurse, *What Is Life?* 62.

12. Dennett, *Kinds of Minds,* 57.

13. Safina, *Becoming Wild,* 43, Kindle.

14. Collingwood, *The Idea of History,* 120.

15. Augustine, *Confessions,* bk. 11, 233.

16. Waldrop, *Complexity,* 330, paraphrasing an interview with Brian Arthur.

17. Chalmers, *What Is This Thing Called Science?* 13.

18. Hume, *Treatise of Human Nature,* 1.3.6.4.

19. Garrett, *Hume,* 17.

20. Wikipedia, s.v. "Maraṇasati," last edited November 4, 2021, 23:49, https://en.wikipedia.org/wiki/Mara%E1%B9%87asati.

21. I borrowed this example from Pinker, *How the Mind Works*, 106–7.

22. Cited in Silver, *Signal and the Noise*, 230.

23. Nate Silver's 2012 book on prediction is called *Signal and the Noise*.

24. Rescher, *Predicting the Future*, 61.

25. Rescher, "Predicting and Knowability," 118.

26. Ord, *The Precipice*, 79.

27. Silver, *Signal and the Noise*, 61.

28. Vikram Mansharamani, "Navigating Uncertainty: Thinking in Futures," in Schroeter, *After Shock*, 15.

29. Goodwin, *Forewarned*, loc. 127 and 1005, Kindle.

Chapter 3: How Cells Manage the Future

1. Waldrop, *Complexity*, 278.

2. Hume, *Enquiry Concerning Human Understanding*, 4.2.19.

3. Godfrey-Smith, *Metazoa*, loc. 3132, Kindle.

4. LeDoux, *Deep History of Ourselves*, 43; Kant, "Anthropology from a Pragmatic Point of View," cited from de Vito and Della Sala, "Predicting the Future," 1019.

5. Waldrop, *Complexity*, 331.

6. Richerson, "Integrated Bayesian Theory of Phenotypic Flexibility," 54–64.

7. Lyon, "The Cognitive Cell," 4.

8. Lyon, "The Cognitive Cell," 3.

9. Dennett, *Kinds of Minds*, 57.

10. Porter, *Greatest Benefit*, loc. 4430, Kindle; Nurse, *What Is Life?* 9–10.

11. Nurse, *What Is Life?* 10–14.

12. Waldrop, *Complexity*, 278.

13. LeDoux, *Deep History of Ourselves*, 42.

14. Zimmer, *Microcosm*, 146–47.

15. Zimmer, *Microcosm*, 125.

16. Zimmer, *Microcosm*, 113–14.

17. Bray, *Wetware*, loc. 100, Kindle.

18. Nurse, *What Is Life?* 54.

19. Goodsell, *The Machinery of Life*, loc. 198, Kindle; Bray, *Wetware*, loc. 804, Kindle.

20. Bray, *Wetware*, loc. 27, Kindle.

21. The following from Bray, *Wetware*, loc. 775, Kindle; and Roth, *Long Evolution*, 70.

22. Mitchell, *Complexity*, loc. 2445, Kindle.

23. On the flagella, Zimmer, *Microcosm*, 24ff.

Chapter 4: How Plants and Animals Manage the Future

1. Chamovitz, *What a Plant Knows,* loc. 645, Kindle.
2. Mukherjee, *The Gene,* loc. 5797, Kindle.
3. Wolpert, *Developmental Biology,* loc. 1098, Kindle.
4. Nurse, *What Is Life?* 66.
5. Peter Wohlleben's *Hidden Life of Trees* describes the difficult choices trees make during their lives.
6. Chamovitz, *What a Plant Knows,* loc. 248, Kindle.
7. Wolpert, *Developmental Biology,* loc. 682, Kindle.
8. Chamovitz, *What a Plant Knows,* loc. 1033–57, Kindle.
9. Chamovitz, *What a Plant Knows,* chap. 5, specifically loc. 1291ff., Kindle, and see also loc. 548 and 414–98, Kindle; Simard, *Finding the Mother Tree.*
10. On plant "memory," Chamovitz, *What a Plant Knows,* chap. 7, and loc. 906, Kindle.
11. Chamovitz, *What a Plant Knows,* loc. 1684 and 854ff., Kindle; Darwin, *Insectivorous Plants,* from *Works of Charles Darwin,* loc. 96,446, ebook, Mobile Reference.com.
12. Foster and Kreitzman, *Circadian Rhythms,* 108; Chamovitz, *What a Plant Knows,* loc. 261, Kindle.
13. Chamovitz, *What a Plant Knows,* loc. 1795, Kindle.
14. Foster and Kreitzman, *Circadian Rhythms,* xvii, 11, 45.
15. My paraphrase, based on Foster and Kreitzman, *Circadian Rhythms,* 1.
16. Foster and Kreitzman, *Circadian Rhythms,* 57; see 125–27 on the simplest possible circadian clocks.
17. Darwin, *Power of Motion in Plants,* from *Works of Charles Darwin,* loc. 105,592–105,607, ebook, MobileReference.com; Peter Wohlleben flirts with the same idea in *Hidden Life of Trees,* 62.
18. Darwin, *Power of Movement in Plants,* from *Works of Charles Darwin,* loc. 98,428, ebook, MobileReference.com.
19. Chamovitz, *What a Plant Knows,* loc. 375, Kindle.
20. Sheldrake, *Entangled Life,* chap. 4. My thanks to Robin Christian for this reference.
21. Sabrin et al., "Hourglass Organization of the C. elegans Connectome."
22. Churchland, *Braintrust,* 44, citing Rodolfo Llinás, *I of the Vortex: From Neurons to the Self* (Cambridge, MA: MIT Press, 2002).
23. Roth, *Long Evolution,* 82, chap. 7.
24. LeDoux, *Deep History,* 112, 137; Roth, *Long Evolution,* 79ff., chap. 7.
25. Roth, *Long Evolution,* 98.
26. Roth, *Long Evolution,* 94, 115.
27. Davies, *Demon in the Machine,* 195.
28. O'Shea, *The Brain,* 131, and see 52.
29. Roth, *Long Evolution,* 234, 226.
30. Roth, *Long Evolution,* chap. 5.
31. From Kandel, *In Search of Memory,* loc. 1243, Kindle.
32. LeDoux, *The Deep History,* 61.

33. O'Shea, *The Brain,* 31; Roth, *Long Evolution,* 67.
34. Based on O'Shea, *The Brain,* chap. 3.
35. Kandel, *In Search of Memory,* loc. 1449, Kindle.
36. Kandel, *In Search of Memory,* loc. 1195, Kindle.
37. Kandel, *In Search of Memory.*
38. Kandel, *In Search of Memory,* loc. 3518, 3146, 3844, Kindle.
39. LeDoux, *Deep History,* 31.
40. Kandel, *In Search of Memory,* loc. 3186, Kindle.
41. Plutarch, *Life of Caesar,* chap. 63.
42. Goodwin, *Forewarned,* loc. 779, Kindle.
43. Gilbert, *Stumbling on Happiness,* 98.
44. Seth, *Being You,* 96–101, Kindle.
45. Kahneman, *Thinking, Fast and Slow.*
46. Kahneman, *Thinking, Fast and Slow,* chap. 10; "facetiously" because, as we will see in chapter 7, you should only trust statistical conclusions if they are based on "large numbers."
47. Kahneman, *Thinking, Fast and Slow,* 25.
48. Russell, *Human Compatible,* 16.
49. Gopnik, *The Philosophical Baby,* 119; the latest survey is Seth, *Being You.*

Chapter 5: What Is New about Human Future Thinking?

1. Wordsworth and Wordsworth, *Penguin Book of Romantic Poetry,* 255.
2. Sornette, "Dragon-kings."
3. Roth, *Long Evolution,* 251.
4. Safina, *Becoming Wild,* 59, on sperm whale brains; Roth, *Long Evolution,* 232 and table on 226.
5. Churchland, *Conscience,* 24.
6. Dunbar, *Human Evolution.*
7. How complex and stressful these calculations can be is apparent from Cheney and Seyfarth, *Baboon Metaphysics.*
8. Roth, *Long Evolution,* 234, 260; Churchland, *Braintrust,* 119.
9. Kahneman, *Thinking, Fast and Slow.*
10. Safina, *Becoming Wild,* is a beautiful account of the richness of animal cultures; on "collective learning," Christian, *Maps of Time* and *Origin Story;* on cultural evolution, Mesoudi, *Cultural Evolution,* is a useful introduction.
11. Mesoudi, *Cultural Evolution,* 203.
12. Steven Pinker, *The Language Instinct,* chap. 1, loc. 115, Kindle.
13. The role of cooperation is stressed in works by Michael Tomasello, such as *Why We Cooperate.*
14. Roth, *Long Evolution,* 260.
15. Goswami, *Child Psychology,* 52.
16. Gopnik, *The Philosophical Baby,* 28.
17. On education in oral cultures, see Kelly, *Knowledge and Power,* 31–32; Karl Popper, from Plotkin, *Darwin Machines,* 69–70: "the growth of our

knowledge is the result of a process closely resembling what Darwin called 'natural selection.'"

18. Ferguson, *Essay on the History of Civil Society,* 7.

19. Whitehead, *Adventures of Ideas,* 93, in a chapter on foresight.

20. Noble and Davidson, "Tracing the Emergence," for some stern cautions about such methodologies.

21. From Gell, *The Anthropology of Time,* 126.

22. Eliade, *Myth of the Eternal Return.*

23. Gell, *The Anthropology of Time,* 127.

24. From Goody, "Time: Social Organization," 31.

25. Gell, *The Anthropology of Time,* 315.

26. Goody, "Time: Social Organization," 31.

27. Goody, "Time: Social Organization."

28. Elias, *Time,* 144.

29. Gell, *The Anthropology of Time,* 3.

30. Goody, "Time: Social Organization," 30.

31. Christian, *Maps of Time,* 254 and 209.

32. Rose, *Dingo Makes Us Human,* 5.

33. A central argument of Kelly, *Knowledge and Power,* see chap. 2.

34. Marshall Thomas, *The Old Way,* 266.

35. Haynes, "Astronomy and the Dreaming," 54.

36. Marshack, *The Roots of Civilization;* reasons for skepticism are listed in Noble and Davidson, "Tracing the Emergence," 127–29.

37. Sahlins, "Original Affluent Society," 22, citing Lee and DeVore, *Man the Hunter,* 37.

38. Kelly, *Knowledge and Power,* 133.

39. Haynes, "Astronomy and the Dreaming," 54.

40. McGrath and Jebb, *Long History, Deep Time,* 4.

41. A central argument of Swain, *A Place for Strangers.*

42. Goody, "Time: Social Organization," 39.

43. Ecclesiastes 1:4–11 (Authorized [King James] Version).

44. See Sahlins, "Original Affluent Society"; Woodburn, "Egalitarian Societies."

45. Cited in Sahlins, "Original Affluent Society," 27.

46. Kelly, *Knowledge and Power,* 117.

47. Cicero, *On the Nature of the Gods,* bk. 1, 3–5.

48. Goswami, *Child Psychology,* 34–35.

49. There is now a discipline devoted to the cognitive science of religion: see Guthrie, *Faces in the Clouds;* Boyer, *Religion Explained;* Larson, *Understanding Greek Religion.*

50. Larson, *Understanding Greek Religion,* 74–75.

51. Rawson, *Cicero,* 241.

52. Marshall Thomas, *The Old Way,* 261.

53. The following from Marshall Thomas, *The Old Way,* 269–73.

54. Lewis-Williams, *Conceiving God,* loc. 4604, Kindle.

Chapter 6: Future Thinking in the Agrarian Era

1. From Johnston, *Ancient Greek Divination,* 7–8.
2. Data from Our World in Data entries on population, deforestation, and urbanization; and from Christian, *Origin Story,* 312, based mainly on Smil, *Harvesting the Biosphere.*
3. Richerson, Boyd, and Bettinger, "Was Agriculture Impossible?"
4. Genesis 8:15–17, 9:2 (Authorized [King James] Version).
5. Goody, "Time: Social Organization," 39–41; *Epic of Gilgamesh,* accessed July 3, 2021, http://www.ancienttexts.org/library/mesopotamian/gilgamesh /tab1.htm.
6. Cicero, *On Divination,* bk. 2, 383, 379; strictly, these opinions are those of the "character" of Cicero in dialogue with his brother, Quintus, which is why it is hard to be certain about Cicero's real views.
7. Revelation 4:1 (Authorized [King James] Version).
8. Hobbes, *Leviathan,* chap. 12, "Of Religion."
9. Jaspers, *Origin and Goal of History,* and see Eisenstadt, "Axial Age."
10. From Bellah, *Religion in Human Evolution,* 268.
11. The phrase "imagined communities" is borrowed from Benedict Anderson's classic study of nationalism, *Imagined Communities.*
12. Cicero, *On the Laws,* 411; citation from Cicero, *On Divination,* bk. 2, 451.
13. Atwood, *Encyclopedia of Mongolia and the Mongol Empire,* 494–95.
14. From Christian, *History of Russia, Central Asia and Mongolia,* 1:59–61.
15. De Rachewiltz, *Secret History of the Mongols,* 1:457–60; other translations are possible; Atwood, *Encyclopedia of Mongolia and the Mongol Empire,* 99.
16. From Christian, *History of Russia, Central Asia and Mongolia,* 1:425.
17. De Rachewiltz, *Secret History of the Mongols,* secs. 244–46 (1:168–74).
18. Atwood, *Encyclopedia of Mongolia and the Mongol Empire,* 100.
19. From Christian, *History of Russia, Central Asia and Mongolia,* 1:425.
20. Thomas and Humphrey, *Shamanism, History and the State,* 11.
21. Johnston, *Ancient Greek Divination,* 3.
22. From Raphals, *Divination,* 253, in turn from Xenophon, *Anabasis,* 4.3.17–19.
23. Johnston, *Ancient Greek Divination,* 11–12; Beard, "Cicero and Divination: The Formation of a Latin Discourse," 33–46; most scholars take Cicero's skepticism more seriously.
24. Flower, *Seer in Ancient Greece,* 34.
25. Johnston, *Ancient Greek Divination,* 33–36.
26. Johnston, *Ancient Greek Divination,* 49.
27. Hobbes, *Leviathan,* chap. 12, "Of Religion"; Strathern, *Brief History of the Future,* 13.
28. Johnston, *Ancient Greek Divination,* 69–70.
29. Parke and Wormell, *The Delphic Oracle,* 1:189.
30. Raphals, *Divination,* 220.
31. Nissinen, Ritner, and Seow, *Prophets and Prophecy,* 25.
32. Raphals, *Divination,* 148; Flower, *The Greek Seer,* 32.

33. Flower, *The Greek Seer*, 32–34.
34. Raphals, *Divination*, 72.
35. Keightley, "The Shang," 247, 252.
36. Raphals, *Divination*, 43; Keightley, "The Shang," 236–37.
37. Keightley, *These Bones Shall Rise Again*, 102.
38. Raphals, *Divination*, 88–89.
39. Keightley, "The Shang," 236–37.
40. Keightley, *These Bones Shall Rise Again*, 103.
41. Keightley, *These Bones Shall Rise Again*, 127.
42. Keightley, *These Bones Shall Rise Again*, 129.
43. Keightley, *These Bones Shall Rise Again*, 130; Raphals, *Divination*, 182–83.
44. Raphals, *Divination*, 205.
45. Raphals, *Divination*, 165.
46. Keightley, "The Shang," 256, and *These Bones Shall Rise Again*, 109.
47. Campion, *Astrology and Cosmology*, for a global overview.
48. *King Lear*, act 1, scene 2.
49. Raphals, *Divination*, 136.
50. Pankenier, *Astrology and Astronomy in Early China*, 6–7.
51. Raphals, *Divination*, 136.
52. For a modern translation of the ancient form of the *I Ching* with lengthy commentary, see Redmond, *The I Ching*.
53. Karl Jung, 1949, from Redmond, *The I Ching*, 22.
54. Translation from Redmond, *The I Ching*, 63.
55. Keightley, "The Shang," 258–60.
56. Raphals, *Divination*, 94, quotation from 99.
57. Bacigalupo, *Shamans of the Foye Tree*, 17.
58. Lewin, "Popular Religion," 68.
59. Lewin, "Popular Religion," 64; Ryan, *Bathhouse at Midnight*, 51–52.
60. Ryan, *Bathhouse at Midnight*, 44.
61. Ryan, *Bathhouse at Midnight*, 96, 100, 108.
62. Evans-Pritchard, *Witchcraft*, 142–43.
63. From Christian, *History of Russia, Central Asia and Mongolia*, 2:343–44.
64. From Christian, *History of Russia, Central Asia and Mongolia*, 1:59.
65. Tedlock, "Toward a Theory of Divinatory Practice," 65.
66. Bacigalupo, *Shamans of the Foye Tree*, 26.
67. Cicero, *On Divination*, bk. 1, 297, 345; Johnston, *Ancient Greek Divination*, 9.
68. Vitebsky, *The Shaman*, 112–13.
69. Evans-Pritchard, *Witchcraft*, 79–80.
70. Evans-Pritchard, *Witchcraft*, 73.
71. Evans-Pritchard, *Witchcraft*, 102–7.
72. Cicero, *On Divination*, bk. 1, 369.
73. Augustine, *Confessions*, bk. 7, 117.
74. Evans-Pritchard, *Witchcraft*, 108–9.
75. Beard, *SPQR*, 465; the following is based on chap. 10 of Hansen, *Anthology of Ancient Greek Popular Literature*, which translates what scholars call the "second edition" of the oracle.

76. A later version added questions for soldiers and farmers; Stewart, *Sortes Barberinianae,* 185–88.

77. Toner, *Popular Culture in Ancient Rome,* 48.

78. Luijendijk and Klingshirn, *My Lots Are in Thy Hands,* 1; Hansen, *Anthology of Ancient Greek Popular Literature,* 285–86.

79. Toner, *Popular Culture in Ancient Rome,* 47–48.

Chapter 7: Modern Future Thinking

1. Lukes and Urbinati, *Condorcet,* 125.

2. Steffen et al., "Trajectory of the Anthropocene."

3. Data in this paragraph from Our World in Data website and from Christian, *Origin Story,* 312, based mainly on Smil, *Harvesting the Biosphere.*

4. Such as Arthur, *The Nature of Technology,* and Headrick, *Humans versus Nature.*

5. For a big-history perspective on globalization, see Christian, foreword and introduction to Zinkina et al., *Big History of Globalization.*

6. Ogle, *Global Transformation of Time,* 1–2.

7. Whitehead, *Adventures of Ideas,* 93, in a chapter on foresight.

8. From Fernandez-Armesto, *The World,* CD.

9. On the discovery of deep time, Toulmin and Goodfield, *The Discovery of Time,* though dated, is very good.

10. On the chronometric revolution, see Christian, "History and Science after the Chronometric Revolution."

11. Shapin, *The Scientific Revolution,* 10; quotation from 37.

12. Davies, *Magic,* 45; On the borrowing of Schiller's phrase "the disenchantment of the world," see Gerth and Mills, *From Max Weber,* 51.

13. "Science as a Vocation," in Gerth and Mills, *From Max Weber,* 139.

14. Shapin, *The Scientific Revolution,* 154 and 33.

15. Paraphrase of Wooton, *The Invention of Science,* 5–6, 8–9.

16. Porter, *Greatest Benefit,* loc. 8991, Kindle.

17. As Shapin notes, this may have been just a thought experiment: *The Scientific Revolution,* 84; on Torricelli, see Dewdney, *Epic Drama,* 152ff.

18. Silver, *Signal and the Noise,* 372.

19. Pearl, "Art and Science of Cause and Effect," 415.

20. From Weaver, *Lady Luck,* 74.

21. Gilmour, "Nature and Function of Astragalus Bones."

22. Stewart, *Do Dice Play God?* 28; Mlodinow, *Drunkard's Walk,* loc. 806ff., Kindle. My thanks to Nic Baker for giving me access to his research on Cardano.

23. Cited in Mlodinow, *The Drunkard's Walk,* loc. 1007 and 1064, Kindle.

24. Stewart, *Do Dice Play God?* 29–30.

25. Mlodinow, *The Drunkard's Walk,* chap. 3; the idea of a sample space is described rigorously in the first chapter of the classic *Introduction to Probability Theory* by William Feller.

26. Stewart, *Do Dice Play God?* 43–44; experiments have shown that even tosses of ordinary coins do not give perfectly random results.

27. This discussion is based partly on Mlodinow, *The Drunkard's Walk*, loc. 1064ff., Kindle, though he attributes this finding to Galileo.

28. Daston, *Classical Probability*, 15.

29. Weaver, *Lady Luck*, 42.

30. Pascal, *Pensées*, in the section beginning: "Infinite — nothing."

31. Arnauld, et al., *Logic, or the Art of Thinking*, 274–75.

32. Stewart, *Do Dice Play God?* 33 and 91.

33. McGrayne, *Theory That Would Not Die*, 7.

34. Rosenbaum, "100 Years of Heights and Weights," 281, summary data from 282.

35. Hume cited in Hacking, *The Taming of Chance*, 13; Daston, *Classical Probability*, 10.

36. Isaacson, *Einstein*, 325.

37. Hacking, *The Emergence of Probability*, 105–6.

38. Hacking, *The Taming of Chance*, 40.

39. From Hacking, *The Taming of Chance*, 41.

40. Hacking, *The Taming of Chance*, 2–3 and passim.

41. Hacking, *The Taming of Chance*, 105.

42. Mayer-Schönberger and Cukier, *Big Data*, 6.

43. Mayer-Schönberger and Cukier, *Big Data*, 11.

44. Urry, *What Is the Future?* 89.

45. Holmes, *Big Data*, 27.

46. Bell, *Foundations of Futures Studies*, 1:44,

47. Meadows, Meadows, and Randers, *Beyond the Limits*, 199.

48. See Turner, "Comparison of the Limits to Growth," and "Is Global Collapse Imminent?"; Herrington, "Update to Limits to Growth," 2021.

49. Ord, *The Precipice*, 70–73.

50. Dewdney, *Epic Drama*, 154–58.

51. Blum, *The Weather Machine*, 19–28; Dewdney, *Epic Drama*, 158ff.

52. Blum, *The Weather Machine*, 125; on the ECMWF, see Blum, *The Weather Machine*, chap. 8.

53. Silver, *Signal and the Noise*, 181–82.

54. For brief overviews, see Gidley, *The Future*, 58; Sardar, *Future: All That Matters*, chap. 3; and Bell, *Foundations of Futures Studies*, 1, chap. 1.

55. On Wells as a founder of modern futures studies, see Wagar, "H.G. Wells and the Genesis of Future Studies."

56. Wells, "Discovery of the Future," 1902.

57. From Sardar, *Future: All That Matters*, loc. 350, Kindle.

58. Figures from Bell, *Foundations of Futures Studies*, 1:63–64.

59. Strathern, *Brief History of the Future*, 205ff. and 263ff.

60. Andersson, *Future of the World*, 4; on Flechtheim, see Strathern, *Brief History of the Future*, chap. 4.

61. Gidley, *The Future*, 5–6, 51.

62. Meadows et al., *The Limits to Growth*, loc. 398–414, Kindle.

63. Gidley, *The Future*, 55–56; on the WFSF, see https://wfsf.org/; Sardar, *Future: All That Matters*, loc. 461ff., Kindle, discusses diverse visions of futures

studies from different regional perspectives; there is now an association for professional futurists (https://www.apf.org/), whose website says it has four hundred members from forty countries.

64. Bell, *Foundations of Futures Studies,* vol. 1, chap. 2, is on "Purposes of Future Studies"; see also 1:102–12.

65. Bell, *Foundations of Futures Studies;* and Aligica, "Special Edition on Wendell Bell; textbook introductions to the techniques of professional futurists include Hines and Bishop, *Thinking about the Future;* on scenario planning, Schwartz, *The Art of the Long View.*

66. On the relationship between science fiction and futures studies, see James and Mendlesohn, "Fiction and the Future."

Chapter 8: Near Futures

1. Rees, *On the Future,* 12.

2. From Krznaric, *The Good Ancestor,* 89.

3. Dator, *A Noticer in Time,* 42; Harman, *Incomplete Guide to the Future,* from Joe Voros, "Philosophical Foundations," 69.

4. Maslow, "Theory of Human Motivation" and "Symposium: Revisiting Maslow."

5. Christian, "History and Global Identity."

6. The declaration can be found at "Towards a Global Ethic," 1993, Parliament of the World's Religions, https://parliamentofreligions.org/towards -global-ethic-initial-declaration.

7. From Sargent, *Utopianism,* 15.

8. Sargent, *Utopianism,* 66, citing an 1833 English translation of a work by a Catholic priest, Father Sangermano.

9. Lukes and Urbinati, *Condorcet,* 7.

10. Lukes and Urbinati, *Condorcet,* 126, 45, 96.

11. Lukes and Urbinati, *Condorcet,* 136, 145.

12. On Wells's influence, see Hensel, "H.G. Wells and the Drafting of a Universal Declaration of Human Rights"; and see Wells, *Rights of Man; or, What Are We Fighting For?* 23.

13. Lukes and Urbinati, *Condorcet,* 136–37.

14. Malthus, *Essay on the Principle of Population.*

15. Malthus, *Essay on the Principle of Population,* 18–20.

16. From Bell, *Foundations of Futures Studies,* 1:117.

17. Meadows et al., *The Limits to Growth.*

18. Meadows et al., *The Limits to Growth,* 196. The phrase "Copernican revolution of the mind" was added by the project's Club of Rome sponsors.

19. Meadows et al., *The Limits to Growth,* 24–25.

20. From Meadows et al., *The Limits to Growth,* 175.

21. From "United Nations Framework Convention on Climate Change," United Nations, 1992, https://unfccc.int/files/essential_background/back ground_publications_htmlpdf/application/pdf/conveng.pdf.

Notes

22. *1992 World Scientists' Warning to Humanity,* July 16, 1992, Union of Concerned Scientists, https://www.ucsusa.org/resources/1992-world-scientists-warning-humanity.

23. "The Future We Want," United Nations General Assembly, September 11, 2012, https://sustainabledevelopment.un.org/index.php?page=view&type=400&nr=733&menu=35.

24. Preamble to the UN Sustainable Development Goals of 2015, from "Transforming Our World: The 2030 Agenda for Sustainable Development," United Nations General Assembly, October 21, 2015.

25. The Sustainable Goals in their latest form can be downloaded from "Transforming Our World: The 2030 Agenda for Sustainable Development," United Nations, 2015, https://sustainabledevelopment.un.org/post2015/transformingourworld/publication.

26. Randers, *2052,* loc. 1319, Kindle.

27. Rumsfeld's press conference can be downloaded from https://archive.ph/20180320091111/http://archive.defense.gov/Transcripts/Transcript.aspx?TranscriptID=2636; Silver, *Signal and the Noise,* 420–21, for a good discussion.

28. From Raworth, *Doughnut Economics,* 124.

29. Taleb, *The Black Swan.*

30. Eldredge and Gould, "Punctuated Equilibria."

31. I tell this story in Christian, *Origin Story.*

32. The following figures are from Kaku, *Physics of the Future,* 328ff.

33. Rosling and Rosling, *Factfulness,* 51.

34. Max Roser, Hannah Ritchie, and Bernadeta Dadonaite, "Child and Infant Mortality," Our World in Data, last updated November 2019, https://ourworldindata.org/child-mortality; Smil, *Numbers Don't Lie,* 9.

35. Holmes, *The Age of Wonder,* 305ff., summarizes her horrifying account; Porter, *Greatest Benefit,* loc. 7145, Kindle, gives extracts.

36. Pinker, *Better Angels of Our Nature;* for a summary of Pinker's argument, see Steven Pinker, "A History of Violence: Edge Master Class 2011," Edge, September 27, 2011, https://www.edge.org/conversation/mc2011-history-violence-pinker.

37. Schwab, *Stakeholder Capitalism,* 25.

38. The crucial graphs can be found at Max Roser, "Future Population Growth," Our World in Data, last revised November 2019, https://ourworldindata.org/future-population-growth; Ehrlich and Ehrlich, *The Population Bomb.*

39. Vollset et al., "Fertility, Mortality, Migration, and Population Scenarios."

40. Data from Max Roser, "Fertility Rate: Children Born per Woman [World]," Our World in Data, first published in 2014; substantive revision published December 2, 2017, https://ourworldindata.org/fertility-rate.

41. Weart, "Development of the Concept of Dangerous Anthropogenic Climate Change."

42. Riahi et al., "Shared Socioeconomic Pathways," explains the underlying approach.

43. Allan et al., *Climate Change 2021,* 16–18.

44. Allan et al., *Climate Change 2021*, 36.

45. Figures from Bar-On, Phillips, and Milo, "Biomass Distribution on Earth."

46. S. Díaz et al., IPBES Global Assessment (2019), 24.

47. The 2021 UNEP report, *Making Peace with Nature*, gives some terrifying statistics.

48. Lovelock, *Gaia: A New Look at Life on Earth*.

49. Rockström and Klum, *Big World: Small Planet*.

50. From Raworth, *Doughnut Economics*, 47.

51. There is a good discussion of the odds in Ord, *The Precipice*, 24–26 and 90–102.

52. Piketty, *Capital in the Twenty-First Century*.

53. Scheidel, *The Great Leveler*, introduction.

54. Scheidel, *The Great Leveler*.

55. Al-Khalili, *What's Next?* surveys new technologies already on the drawing board.

56. Christian, "The Noösphere."

57. Randers, *2052*, loc. 670, Kindle.

58. Dator, *A Noticer in Time*, chap. 5, pt. 4, "The Four Generic Futures."

59. Ord, *The Precipice*, 37, for the definition of an existential catastrophe; 167, for the table of likely existential crises.

60. Greer's *The Long Descent* describes such a slow descent.

61. Greer, *The Long Descent*, 83.

62. Raskin, *Journey to Earthland*.

63. Sinclair and LaPlante, *Lifespan*.

64. Meadows et al., *Limits to Growth*, 171.

65. On transhumanism, see Grinin and Grinin, "Crossing the Threshold of Cyborgization."

66. From Schwab, *Stakeholder Capitalism*, 167.

Chapter 9: Middle Futures

1. Ord, *The Precipice*, 52.

2. Wells, *The Outline of History*, vol. 2, chap. 41, pt. 4.

3. Such as Wagar, *Short History of the Future*, and Stableford and Langford, *The Third Millennium*.

4. Gates, *How to Avoid a Climate Disaster*, chap. 4.

5. Kaku, *Physics of the Future*, 252.

6. Kaku, *Physics of the Future*, 246.

7. Kardashev, "Transmission of Information by Extra-Terrestrial Civilizations" and "On the Inevitability and the Possible Structures of Supercivilizations."

8. Sagan, *The Cosmic Connection*, chap. 34.

9. Rorvig, "How to Spot an Alien Megastructure," on recent attempts to detect such structures.

10. Voros, "Big Futures," 423, on Hoag's objects; Kaku, *Physics of the Future*, 330.

11. Feynman, "There's Plenty of Room at the Bottom."
12. John Smart, "Exponential Progress," in Schroeter, *After Shock*, 499.
13. This is the vision of Drexler, *Radical Abundance*.
14. On AI, see Russell, *Human Compatible*.
15. Kaku, *Future of Humanity*, 125.
16. Bostrom, *Superintelligence*, 123.
17. Kaku, *Physics of the Future*, 109.
18. Strathern, *Brief History of the Future*, 296.
19. Srubar, "Buildings Grown by Bacteria."
20. Natasha Vita-More, "A History of Transhumanism," in Lee, *The Transhumanism Handbook*, chap. 2, 49.
21. Lem, "Thirty Years Later," from Swirski, *Art and Science of Stanislaw Lem*.
22. Gerjuoy, "Most Significant Events of the Next Thousand Years."
23. Finney, *From Sea to Space*, chap. 3, "One Species or a Million?" 105.
24. Finney, *From Sea to Space*, chap. 3, "One Species or a Million?" 113.
25. From Caitlin Yilek, "Jeff Bezos on Future of Spaceflight," CBS News, July 21, 2021, https://www.cbsnews.com/news/jeff-bezos-space-heavy-industry-polluting-industry/.
26. Finney, *From Sea to Space*, 105.
27. Kaku, *Future of Humanity*, 107.
28. Ord, *The Precipice*, 231.
29. There is a good account of the evolution of astrobiology and planetary science in Grinspoon, *Earth in Human Hands*.
30. Olaf Stapledon, *Starmaker*, cited in Kaku, *Future of Humanity*, 244.
31. Shostak, "The Value of 'L,'" 404.
32. Robinson, "Realism of Our Times."
33. Kaku, *Physics of the Future*, 340; Dyson, "Time without End," 453.

Chapter 10: Remote Futures

1. Jacob Margolis (@JacobMargolis), "My battery is low and it's getting dark," Twitter, February 12, 2019, 4:38 p.m., https://twitter.com/jacobmargolis/status/1095436913173880832.
2. Bar-On, Phillips, and Milo, "Biomass Distribution on Earth."
3. The following is based in part on Meadows, *Future of the Universe*, chap. 2.
4. See Robin George Andrews, "Watch This Billion-Year Journey of Earth's Tectonic Plates," *New York Times*, February 6, 2021, https://www.nytimes.com/2021/02/06/science/tectonic-plates-continental-drift.html, for a film of plate tectonics over the last billion years.
5. Nance et al., "The Supercontinent Cycle."
6. Meadows, *Future of the Universe*, 111.
7. Meadows, *Future of the Universe*, 117 and 114.
8. In *Extraterrestrial*, Avi Loeb, head of the astronomy department at Harvard, speculates that it may have been made by intelligent beings, but few others take this idea seriously.
9. Meadows, *Future of the Universe*, chap. 2.

10. Voros, "Big Futures," 417.

11. Meadows, *Future of the Universe*, 18.

12. Lovelock, *Gaia*.

13. Shah, "Complex Life's Days Are Numbered."

14. Meadows, *Future of the Universe*, 65–66.

15. Meadows, *Future of the Universe*, 126.

16. On galaxy collisions, Meadows, *Future of the Universe*, chap. 10; on the collision with Andromeda, Mack, *The End of Everything*, 50–51.

17. Walls, *Oxford Handbook to Eschatology*, 151.

18. Dyson, "Time without End."

19. Walls, *Oxford Handbook to Eschatology*, 3.

20. Walls, *Oxford Handbook to Eschatology*, 6.

21. Klee "Spiritualism: The Technological Endgame," in Schroeter, *After Shock*, 65.

22. Mack, *The End of Everything*, 61.

23. Hawking, *Brief History of Time*, 150–51.

24. Mack, *The End of Everything*, 95.

25. Holt, *When Einstein Walked with Gödel*, 18.

26. Holt, *When Einstein Walked with Gödel*, 243.

27. Smolin, *Life of the Cosmos*.

28. Rees, *Just Six Numbers*.

29. Meadows, *Future of the Universe*, 162.

Bibliography of Cited Sources

Note: I have only cited sources used directly.

Aligica, Paul Dragos, ed. "Special Issue on Wendell Bell." *Futures* 43, no. 6 (2011): 563–638.

Al-Khalili, Jim, ed. *What's Next: Even Scientists Can't Predict the Future — or Can They?* London: Profile, 2017.

Allan, Richard P., P. A. Arias, S. Berger, J. G. Canadell, C. Cassou, D. Chen, A. Cherchi, et al., eds. *Climate Change 2021: The Physical Basis: Summary for Policy Makers.* Cambridge: Cambridge University Press, 2021.

Anderson, Benedict. *Imagined Communities: Reflections on the Origins and Spread of Nationalism.* 1983. Rev. ed., London: Verso, 2016, with the preface to the 1991 ed.

Anderson, P. W. "More Is Different: Broken Symmetry and the Hierarchical Structure of Science." *Science* 177, no. 4047 (1972): 393–96.

Andersson, Jenny. *The Future of the World: Futurology, Futurists, and the Struggle for the Post Cold War Imagination.* Oxford: Oxford University Press, 2018.

Arnauld, Antoine, and Pierre Nicole. *Logic, or the Art of Thinking.* Translated by Jill Vance Buroker. Cambridge: Cambridge University Press, 1996.

Arthur, Brian. *The Nature of Technology.* New York: Penguin, 2009.

Asimov, Isaac. *Foundation* (1951), *Foundation and Empire* (1952), and *Second Foundation* (1953). New York City: Gnome Press.

Atwood, Christopher P. *Encyclopedia of Mongolia and the Mongol Empire.* New York: Facts on File, 2004.

Augustine. *City of God.* Bks. 1–22. Loeb Classical Library. Cambridge, MA: Harvard University Press, 1957, 411–417.

———. *Confessions.* Translated by Henry Chadwick. New York: Oxford University Press, 1992.

Bacigalupo, Ana Mariella. *Shamans of the Foye Tree: Gender, Power, and Healing among Chilean Mapuche.* Austin: University of Texas Press, 2010. My thanks to Merry Wiesner-Hanks for this reference.

Bardon, Adrian. *A Brief History of the Philosophy of Time.* New York: Oxford University Press, 2013.

Baron, Sam, and Kristie Miller. *An Introduction to the Philosophy of Time.* Cambridge: Polity Press, 2019.

Bar-On, Yinon M., Rob Phillips, and Ron Milo. "The Biomass Distribution on Earth." *Proceedings of the National Academy of Science* 115, no. 25 (2018): 6506–11.

Beard, Mary. "Cicero and Divination: The Formation of a Latin Discourse." *Journal of Roman Studies* 76 (1986): 33–46.

———. *SPQR: A History of Ancient Rome.* New York: Liveright, 2015.

Bell, Wendell. *Foundations of Futures Studies.* 2 vols. New Brunswick, NJ: Transaction Publishers, 1997, 2004.

Bellah, Robert N. *Religion in Human Evolution: From the Paleolithic to the Axial Age.* Cambridge, MA: Harvard University Press, 2011.

Benjamin, Craig, Esther Quaedackers, and David Baker, eds. *The Routledge Companion to Big History.* London: Routledge, 2020.

Bhagavad Gita. Translated by Laurie L. Patton. London: Penguin, 2014.

Blackburn, Simon. *The Big Questions: Philosophy.* London: Quercus, 2009.

Blum, Andrew. *The Weather Machine: How We See into the Future.* New York: Vintage, 2019.

Boethius. *The Consolation of Philosophy.* Translated by David R. Slavitt. Cambridge, MA: Harvard University Press, 2008.

Bostrom, Nick. *Superintelligence: Paths, Dangers, Strategies.* Oxford: Oxford University Press, 2014.

Boyer, Pascal. *Religion Explained: The Evolutionary Origins of Religious Thought.* Basic Books, 2002.

Bray, Dennis. *Wetwear: A Computer in Every Living Cell.* New Haven, CT: Yale University Press, 2009.

Caldarelli, Guido, and Michele Catanzaro. *Networks: A Very Short Introduction.* Oxford: Oxford University Press, 2012.

Callender, Craig, ed. *The Oxford Handbook of Philosophy of Time.* New York: Oxford University Press, 2011.

Campion, Nicholas. *Astrology and Cosmology in the World's Religions.* New York: New York University Press, 2012.

Carr, E. H. *What Is History?* 1961. Harmondsworth: Penguin, 1964.

Chalmers, Alan F. *What Is This Thing Called Science?* St. Lucia: University of Queensland Press, 1978.

Chamovitz, Daniel. *What a Plant Knows.* 2nd ed. Melbourne: Scribe, 2017.

Cheney, Dorothy L., and Robert M. Seyfarth. *Baboon Metaphysics: The Evolution of a Social Mind.* Chicago: University of Chicago Press, 2007.

Christian, David. "History and Global Identity." In *The Historian's Conscience: Australian Historians on the Ethics of History,* edited by Stuart Macintyre, 139–50. Melbourne: Melbourne University Press, 2004.

———. *A History of Russia, Central Asia and Mongolia.* Vol. 1, *Inner Eurasia from Prehistory to the Mongol Empire.* Oxford: Blackwell, 1998; Vol. 2, *Inner Eurasia from the Mongol Empire to Today, 1260–2000.* Hoboken, NJ: Wiley/Blackwell, 2018.

———. "History and Science after the Chronometric Revolution." In Dick and Lupisella, *Cosmos & Culture,* 441–62.

————. *Maps of Time: An Introduction to Big History.* 2nd ed. Berkeley, CA: University of California Press, 2011.

————. "The Noösphere." In *This Idea Is Brilliant,* edited by John Brockman. New York: Harper Perennial, 2018.

————. *Origin Story: A Big History of Everything.* New York: Little, Brown, 2018.

————. "Silk Roads or Steppe Roads? The Silk Roads in World History." *Journal of World History* 11, no. 1 (2000): 1–26.

Churchland, Patricia. *Braintrust: What Neuroscience Tells Us about Morality.* 2011; Princeton, NJ: Princeton University Press, 2018, with new preface.

————. *Conscience: The Origins of Moral Intuition.* New York: W. W. Norton, 2019.

Cicero. *On the Nature of the Gods. Academics.* Translated by H. Rackham. Loeb Classical Library 268. Cambridge, MA: Harvard University Press, 1933.

————. *On Old Age. On Friendship. On Divination.* Translated by W. A. Falconer. Loeb Classical Library 154. Cambridge, MA: Harvard University Press, 1923.

————. *On the Republic. On the Laws.* Translated by Clinton W. Keyes. Loeb Classical Library 213. Cambridge, MA: Harvard University Press, 1928.

Collingwood, R. G. *The Idea of History.* Rev. ed. Edited by Jan Van Dussen. New York: Oxford University Press, 1994.

Condorcet, Marquis de. *Sketch for a Historical Picture of the Progress of the Human Mind.* In Lukes and Urbinati, *Condorcet: Political Writing,* 1–147.

Cossins, Daniel. "The Time Delusion." *New Scientist,* July 6, 2019, 32–36.

Curd, Martin, and J. A. Cover. *Philosophy of Science: The Central Issues.* New York: W. W. Norton, 1998.

Danks, David. "Safe-and-Substantive Perspectivism." In Massimi and McCoy, *Understanding Perspectivism,* chap. 7.

Darwin, Charles. *Works of Charles Darwin.* MobileReference.com, 2008.

Daston, Lorraine. *Classical Probability in the Enlightenment.* Princeton, NJ: Princeton University Press, 1995. My thanks to Nic Baker for this reference.

Dator, James. *Jim Dator: A Noticer in Time: Selected Work, 1967–2018.* Anticipation Science Book 5. Cham, Switzerland: Springer, 2019.

Davies, Owen. *Magic: A Very Short Introduction.* Oxford: Oxford University Press, 2012.

Davies, Paul. *The Demon in the Machine: How Hidden Webs of Information Are Finally Solving the Mystery of Life.* London: Penguin: 2019.

Dennett, Daniel. *Consciousness Explained.* New York: Penguin, 1991.

————. *Kinds of Minds: Toward an Understanding of Consciousness.* London: Weidenfeld and Nicolson, 1996.

de Rachewiltz, Igor. *The Secret History of the Mongols: A Mongolian Epic Chronicle of the Thirteenth Century.* 2 vols. Leiden: Brill, 2006.

De Vito, Stefania, and Sergio Della Sala. "Predicting the Future." *Cortex* 47, no. 8 (2011): 1018–22.

Dewdney, Christopher. *The Epic Drama of the Atmosphere and Its Weather.* London: Bloomsbury, 2019.

Díaz, S., J. Settele, E. S. Brondízio, H. T. Ngo, M. Guèze, J. Agard, A. Arneth, et al., eds. *IPBES (2019): Summary for Policymakers of the Global Assessment Report on Biodiversity and Ecosystem Services of the Intergovernmental Science-Policy Platform on Biodiversity and Ecosystem Services.* Bonn, Germany: IPBES Secretariat, 2019.

Dick, Steven J., and Mark L. Lupisella, eds. *Cosmos & Culture: Cultural Evolution in a Cosmic Context.* Washington, DC: National Aeronautics and Space Administration, 2009.

Drexler, K. Eric. *Radical Abundance: How a Revolution in Nanotechnology Will Change Civilization.* New York: Perseus, 2013.

Dunbar, Robin. *Human Evolution: A Pelican Introduction.* New York: Penguin, 2014.

Dyson, Freeman. "Time without End: Physics and Biology in an Open Universe." *Reviews of Modern Physics* 51, no. 3 (1979): 447–60.

Egan, Chas A., and Charles H. Lineweaver. "Life, Gravity and the Second Law of Thermodynamics." *Physics of Life Reviews* 5 (2008): 225–42.

Ehrlich, Paul R., and Anne Ehrlich. *The Population Bomb.* New York: Ballantine Books, 1968.

Einstein, Albert. *Relativity, the Special and the General Theory.* Translated by Robert W. Lawson. London: Routledge, 1920.

———. "Zur Elektrodynamik bewegter Körper" ("On the Electrodynamics of Moving Bodies"). *Annalen der Physik* 322 (10): 891–921.

Eisenstadt, Shmuel. "The Axial Age: The Emergence of Transcendental Visions and the Rise of Clerics." *European Journal of Sociology / Archives Européennes de Sociologie Europäisches Archiv für Soziologie* 23, no. 2 (1982): 294–314.

Eldredge, Niles, and Stephen Jay Gould. "Punctuated Equilibria: An Alternative to Phyletic Gradualism." In *Models in Paleobiology,* edited by T. J. M. Schopf, 82–115. San Francisco: Freeman Cooper, 1972.

Eliade, Mircea. *Myth of the Eternal Return, or, Cosmos and History.* Princeton, NJ: Princeton University Press, 1954.

Elias, Norbert. *Time: An Essay.* Oxford: Blackwell, 1992.

Evans-Pritchard, E. E. *Witchcraft, Oracles and Magic among the Azande.* Abridged with an introduction by Eva Gillies. Oxford: Oxford University Press, 1976.

Feller, William. *An Introduction to Probability Theory and Its Applications.* Vol. 1, 3rd ed. New York: John Wiley, 1968.

Ferguson, Adam. *An Essay on the History of Civil Society.* 3rd ed. London, 1768.

Fernandez-Armesto, Felipe. *The World: A History.* Upper Saddle River, NJ: Pearson, 2007.

Feynman, Richard P. *The Character of Physical Law.* 1965. New York: Penguin, 1992.

———. "There's Plenty of Room at the Bottom." Lecture given at the annual meeting of the American Physical Society at the California Institute of Technology, Pasadena, CA, December 29, 1959. http://www.zyvex.com/nanotech/feynman.html.

Finney, Ben. *From Sea to Space.* Auckland, NZ: Massey University Press, 1992.

Flower, Michael A. *The Seer in Ancient Greece*. Berkeley: University of California Press, 2008.

Foster, Russell G., and Leon Kreitzman. *Circadian Rhythms: A Very Short Introduction*. Oxford: Oxford University Press, 2017.

Gallois, William. "Zen History." *Rethinking History* 14, no. 3 (2010): 421–40. https://doi.org/10.1080/13642529.2010.482799.

Garrett, Don. *Hume: The Routledge Philosophers*. New York: Routledge, 2015.

Gates, Bill. *How to Avoid a Climate Disaster*. New York: Penguin, 2021.

Gell, Alfred. *The Anthropology of Time: Cultural Constructions of Temporal Maps and Images*. Oxford: Berg, 1992.

Gerjuoy, Herbert. "The Most Significant Events of the Next Thousand Years." In Slaughter, *Knowledge Base of Futures Studies*, bk. 3, pt. 3.

Gerth, H. H., and C. Wright Mills, ed. *From Max Weber: Essays in Sociology*. London: Taylor & Francis Group, 2013.

Gibelyou, Cameron, and Douglas Northrop. *Big Ideas: A Guide to the History of Everything*. New York: Oxford University Press, 2020.

Gidley, Jennifer M. *The Future: A Very Short Introduction*. Oxford: Oxford University Press, 2017.

Gilbert, Stanley. *Stumbling on Happiness*. London: William Collins, 2007.

Gilmour, G. H. "The Nature and Function of Astragalus Bones from Archaeological Contexts in the Levant and Eastern Mediterranean." *Oxford Journal of Archaeology* 16 (1997): 167–75. Thanks to Ray Laurence for this reference.

Gisin, N. "Mathematical Languages Shape Our Understanding of Time in Physics." *Nature Physics* 16 (2020): 114–16. https://doi.org/10.1038/s415 67-019-0748-5.

Godfrey-Smith, Peter. *Metazoa: Animal Minds and the Birth of Consciousness*. Glasgow, Scotland, William Collins, 2020.

Goodsell, David S. *The Machinery of Life*. New York: Springer, 2009.

Goodwin, Peter. *Forewarned: A Sceptic's Guide to Prediction*. London: Biteback Publishing, 2017.

Goody, Jack. "Time: Social Organization." In *International Encyclopaedia of the Social Sciences*, edited by David Sills, vol. 16, 30–42 (New York: Macmillan, 1968).

Gopnik, Alison. *The Philosophical Baby: What Children's Minds Tell Us about Truth, Love & the Meaning of Life*. New York: Vintage, 2011.

Goswami, Usha. *Child Psychology: A Very Short Introduction*. Oxford: Oxford University Press, 2014.

Greer, John Michael. *The Long Descent: A User's Guide to the End of the Industrial Age*. Gabriola Island, BC, Canada: New Society Publishers, 2008.

Grinin, Anton, and Leonid Grinin. "Crossing the Threshold of Cyborgization." *Journal of Big History* 4, no. 3 (2020): 54–65.

Grinspoon, David. *Earth in Human Hands: Shaping Our Planet's Future*. New York: Grand Central Publishing, 2016.

Guthrie, Stewart. *Faces in the Clouds: A New Theory of Religion*. New York: Oxford University Press, 1993.

Hacking, Ian. *The Emergence of Probability.* 2nd ed. Cambridge: Cambridge University Press, 2006.

———. *The Taming of Chance.* Cambridge: Cambridge University Press, 1990.

Hansen, William. *The Anthology of Ancient Greek Popular Literature.* Bloomington: Indiana University Press, 1998.

Hawking, Stephen. *A Brief History of Time: From the Big Bang to Black Holes.* London: Bantam Press, 1988.

Haynes, Roslynn. "Astronomy and the Dreaming: The Astronomy of the Aboriginal Australians." In *Astronomy across Cultures: The History of non-Western Astronomy,* edited by Helaine Selin. London: Kluwer, 2000.

Headrick, Daniel. *Humans versus Nature: A Global Environmental History.* Oxford: Oxford University Press, 2020.

Hensel, D. Gert. "H.G. Wells and the Drafting of a Universal Declaration of Human Rights." *Peace Research* 35, no. 1 (2003): 93–102.

Herrington, Gaya. "Update to Limits to Growth: Comparing the World3 Model with Empirical Data." *Journal of Industrial Ecology* 24 (2012): 614–26. https://advisory.kpmg.us/articles/2021/limits-to-growth.html.

Hines, Andy, and Peter Bishop. *Thinking about the Future: Guidelines for Strategic Foresight.* 2nd ed. Houston, TX: Hinesight, 2015.

Holmes, Dawn E. *Big Data: A Very Short Introduction.* Oxford: Oxford University Press, 2017.

Holmes, Richard. *The Age of Wonder: How the Romantic Generation Discovered the Beauty and Terror of Science.* Glasgow, Scotland: William Collins, 2008.

Holt, Jim. *When Einstein Walked with Gödel: Excursions to the Edge of Thought.* New York: Farrar, Straus and Giroux, 2018.

Hoyle, Fred. *The Black Cloud.* London: William Heinemann, 1957.

Hume, David. *David Hume Collection* [includes *A Treatise of Human Nature, An Enquiry Concerning Human Understanding, An Enquiry Concerning the Principles of Morals,* and *Dialogues Concerning Natural Religion*]. NP, 2020.

Huxley, Aldous. *Brave New World.* London: Chatto & Windus, 1932.

Isaacson, Walter. *Einstein: His Life and Universe.* New York: Simon & Schuster, 2007.

Ismael, Jenann. *How Physics Makes Us Free.* New York: Oxford University Press, 2016.

———. "Temporal Experience." In *The Oxford Handbook of Philosophy of Time,* edited by Craig Callender, chap. 15, 460–82. New York: Oxford University Press, 2011.

James, Edward, and Farah Mendlesohn. "Fiction and the Future." In Slaughter, *Knowledge Base of Future Studies,* vol. 1, pt. 3.

James, William. *Delphi Complete Works of William James.* East Sussex, UK: Delphi Classics, 2018.

Jaspers, Karl. *Vom Ursprung und Ziel der Geschichte.* 1949. Translated by Michael Bullock as *The Origin and Goal of History.* London: Routledge and Kegan Paul, 1953; citations are from the Routledge Classics edition, 2021.

Johnston, Sarah Iles. *Ancient Greek Divination.* Oxford: Wiley/Blackwell, 2008.

Kahneman, Daniel. *Thinking, Fast and Slow.* New York: Penguin, 2011.

Kaku, Michio. *The Future of Humanity: Terraforming Mars, Interstellar Travel, Immortality, and Our Destiny Beyond.* New York: Penguin, 2018.

———. *Physics of the Future: How Science Will Shape Human Destiny and Our Daily Lives by the Year 2100.* New York: Penguin, 2011.

Kandel, Eric. *In Search of Memory: The Emergence of a New Science of Mind.* New York: W. W. Norton, 2006.

Kant, Immanuel. "Anthropology from a Pragmatic Point of View." Translated by Victor Lyle. Originally published 1798.

Kardashev, N. S. "On the Inevitability and the Possible Structures of Super-civilizations." In *The Search for Extraterrestrial Life: Recent Developments. Proceedings of the 112th Symposium of the International Astronomical Union Held at Boston University, Boston, Mass., U.S.A., June 18–21, 1984,* edited by Michael Papagiannis, 497–504. Dordrecht: D. Reidel, 1985.

———. "Transmission of Information by Extra-Terrestrial Civilizations." *Soviet Astronomy AJ* 8, no. 2 (1964): 217–21. Translated from *Astronomicheskii Zhurnal* 41, no. 2 (1964): 282–87.

Kay, John, and Mervyn King. *Radical Uncertainty: Decision Making for an Uncertain Future.* London: Bridge Street Press, 2020.

Keightley, David N. "The Shang: China's First Historical Dynasty." In *The Cambridge History of Ancient China.* Cambridge: Cambridge University Press, 1999.

———. *These Bones Shall Rise Again: Selected Writings on Early China.* Edited by Henry Rosemont. Albany: State University of New York Press, 2014.

Kelly, Lynne. *Knowledge and Power in Prehistoric Societies: Orality, Memory and the Transmission of Culture.* Cambridge: Cambridge University Press, 2015.

Khayyám, Omar. *The Rubaiyat of Omar Khayyám.* Translated by Edward Fitzgerald. London: Bernard Quaritch, 1859. https://en.wikisource .org/wiki/The_Rubaiyat_of_Omar_Khayyam_(tr._Fitzgerald,_1st _edition).

Kistler, Max. "Causation." In *The Philosophy of Science: A Companion,* edited by Anouk Baberousse, Denis Bonnay, and Mikael Cozic. New York: Oxford University Press, 2018.

Krznaric, Roman. *The Good Ancestor: How to Think Long-Term in a Short-Term World.* London: Penguin, 2020.

Laplace, Pierre-Simon de. "A Philosophical Essay on Probabilities." Translated from the 6th French ed. London: John Wiley, 1902. https://bayes .wustl.edu/Manual/laplace_A_philosophical_essay_on_probabilities .pdf.

Larson, Jennifer. *Understanding Greek Religion.* New York: Routledge, 2016.

LeDoux, Joseph. *The Deep History of Ourselves: The Four-Billion-Year Story of How We Got Conscious Brains.* New York: Viking/Penguin, 2019.

Lee, Newton, ed. *The Transhumanism Handbook.* Cham, Switzerland: Springer, 2019.

Lee, Richard B., and Irven DeVore, eds. *Man the Hunter.* Chicago: Aldine, 1968.

Le Guin, Ursula. *The Dispossessed.* New York: Harper & Row, 1974.

Lewin, Moshe. "Popular Religion in Twentieth Century Russia." In *The Making of the Soviet System: Essays in the Social History of Interwar Russia*, 57–71. London: Methuen, 1985.

Lewis-Williams, David. *Conceiving God: The Cognitive Origin and Evolution of Religion*. London: Thames & Hudson, 2010.

Liu, Cixin. Remembrance of Earth's Past trilogy of novels. New York: Tor, 2006–10.

Loeb, Avi. *Extraterrestrial: The First Sign of Intelligent Lie Beyond Earth*. London: John Murray, 2021.

Lovelock, James. *Gaia: A New Look at Life on Earth*. 1979. Repr., Oxford: Oxford University Press, 1988.

Loy, David. "The Mahāyāna Deconstruction of Time." *Philosophy East and West* 36, no. 1 (1986): 13–23.

Luijendijk, AnneMarie, and William E. Klingshirn, eds. *My Lots Are in Thy Hands: Sortilege and Its Practitioners in Late Antiquity*. Leiden: Brill, 2018.

Lukes, Steven, and Nadia Urbinati, eds. *Condorcet: Political Writing*. Cambridge: Cambridge University Press, 2012, 1–147.

Lyon, Pamela. "The Cognitive Cell: Bacterial Behavior Reconsidered." *Frontiers in Microbiology* 6 (2015). https://doi.org/10.3389/fmicb.2015.00264. My thanks to Martin Robert of Tohoku University for this reference.

Mack, Katie. *The End of Everything (Astrophysically Speaking)*. London: Penguin, 2020.

Malthus, Thomas Robert. *An Essay on the Principle of Population*. Edited by Philip Appleman. New York: W. W. Norton, 1976.

Marshack, Alexander. *The Roots of Civilization: The Cognitive Beginning of Man's First Art, Symbol and Notation*. New York: McGraw-Hill, 1972.

Marshall Thomas, Elizabeth. *The Old Way: A Story of the First People*. New York: Picador, 2006.

Marx, Karl. *The Marx-Engels Reader*. 2nd ed. Edited by Robert C. Tucker. New York: W. W. Norton, 1978.

Maslow, Abraham. "Symposium: Revisiting Maslow: Human Needs in the 21st Century." in *Society* 54 (2017): 508–9. https://doi.org/10.1007/s12115-017-0198-6.

———. "A Theory of Human Motivation." *Psychological Review* 50 (1943): 370–96.

Massimi, Michela, and Casey D. McCoy, eds., *Understanding Perspectivism: Scientific Challenges and Methodological Prospects* (Routledge Studies in the Philosophy of Science). New York: Routledge, 2019.

Mayer-Schönberger, Viktor, and Kenneth Cukier. *Big Data: A Revolution That Will Transform How We Live, Work and Think*. London: John Murray, 2013.

McGrath, Ann. "Deep Histories in Time, or Crossing the Great Divide?" In McGrath and Jebb, *Long History, Deep Time: Deepening Histories of Place*.

McGrath, Ann, and Mary Anne Jebb, eds. *Long History, Deep Time: Deepening Histories of Place*. Canberra: Australian National University Press, 2015.

McGrayne, Sharon B. *The Theory That Would Not Die: How Bayes' Rule Cracked the Enigma Code, Hunted Down Russian Submarines, and Emerged Triumphant from Two Centuries of Controversy.* New Haven, CT: Yale University Press, 2011.

McTaggart, J. Ellis. "The Unreality of Time." *Mind*, n.s., 17, no. 68 (1908): 457–74.

———. "The Unreality of Time" (a restatement of arguments in McTaggart's 1908 article). In *The Philosophy of Time*, edited by Robin Le Poidevin and Murray MacBeath, 23–34. Oxford: Oxford University Press, 1993.

Meadows, A. J. (Jack). *The Future of the Universe.* London: Springer, 2007.

Meadows, D. H., D. L. Meadows, and J. Randers. *Beyond the Limits: Global Collapse or a Sustainable Future.* London: Earthscan, 1992.

Meadows, D. H., D. L. Meadows, J. Randers, and W. W. Behrens. *The Limits to Growth: A Report for the Club of Rome's Project on the Predicament of Mankind.* New York: Universe Books, 1972.

Mellor, D. H. *Real Time.* Cambridge: Cambridge University Press, 1981.

———. *Real Time II.* London: Routledge, 1998.

Mesoudi, Alex. *Cultural Evolution: How Darwinian Theory Can Explain Human Culture and Synthesize the Social Sciences.* Chicago: University of Chicago Press, 2011.

Miller, Walter M. *A Canticle for Leibowitz.* Philadelphia: J. B. Lippincott, 1959.

Mitchell, Melanie. *Complexity: A Guided Tour.* New York: Oxford University Press, 2009.

Mlodinow, Leonard. *The Drunkard's Walk: How Randomness Rules Our Lives.* New York: Pantheon, 2009.

Mukherjee, Siddhartha. *The Gene: An Intimate History.* New York: Scribner, 2016.

Nance, R. Damian, J. Brendan Murphy, and M. Santosh. "The Supercontinent Cycle: A Retrospective Essay." *Gondwana Research* 25 (2014): 4–29.

Neale, Margo. *First Knowledges: The Power and Promise.* Port Melbourne, Victoria, Australia: Thames & Hudson, 2020.

Newton, Isaac. *The Mathematical Principles of Natural Philosophy.* Translated by Andrew Motte. London: Middle-Temple-Gate, 1729.

Nissinen, Martti, Robert Kriech Ritner, and Choon Leong Seow. *Prophets and Prophecy in the Ancient Near East.* Atlanta, GA: Society of Biblical Literature, 2003.

Noble, W., and I. Davidson. "Tracing the Emergence of Modern Human Behavior: Methodological Pitfalls and a Theoretical Path." *Journal of Anthropological Archaeology* 12, no. 2 (1993): 121–49.

Nurse, Paul. *What Is Life? Understand Biology in Five Steps.* Melbourne: Scribe, 2020.

Ogle, Vanessa. *The Global Transformation of Time, 1870–1950.* Cambridge, MA: Harvard University Press, 2015.

Ord, Toby. *The Precipice: Existential Risk and the Future of Humanity.* New York: Hachette, 2020.

O'Shea, Michael. *The Brain: A Very Short Introduction*. Oxford: Oxford University Press, 2005.

Our World in Data. Max Roser et al. https://ourworldindata.org/.

Pankenier, David W. *Astrology and Cosmology in Early China: Conforming Earth to Heaven*. Cambridge: Cambridge University Press, 2013.

Parke, H. W., and D. E. W. Wormell. *The Delphic Oracle*. Vol. 1, *The History*. Vol. 2, *The Oracular Responses*. Oxford: Blackwell, 1956.

Pearl, Judea. "The Art and Science of Cause and Effect." Public lecture, UCLA Faculty Research Lectureship Program, 1996. Reprinted as the epilogue to Pearl, *Causality: Models, Reasoning, and Inference*. New York: Cambridge University Press, 2009, 401–28. http://bayes.cs.ucla.edu/BOOK-2K/causality2-epilogue.pdf.

Pearl, Judea, and Dana Mackenzie. *The Book of Why: The New Science of Cause and Effect*. London: Penguin, 2018.

Piketty, Thomas. *Capital in the Twenty-First Century*. Translated by Arthur Goldhammer. Cambridge, MA: Harvard University Press, 2014.

Pinker, Steven. *The Better Angels of Our Nature: Why Violence Has Declined*. New York: Viking, 2011.

––––––. *How the Mind Works*. London: Allen Lane, 1998.

––––––. *The Language Instinct: How the Mind Creates Language*. New ed. London: Penguin, 2003.

Plotkin, Henry. *Darwin Machines and the Nature of Knowledge*. Cambridge, MA: Harvard University Press, 1994.

Plutarch. *Life of Caesar*. In *Lives*, vol. 7, *Demosthenes and Cicero. Alexander and Caesar*. Translated by Bernadotte Perrin. Loeb Classical Library 99. Cambridge, MA: Harvard University Press, 1919.

Polak, Fred. *The Image of the Future*. Translated by Elise Boulding. Amsterdam: Elsevier, 1973.

Porter, Roy. *The Greatest Benefit to Mankind: A Medical History of Humanity*. Glasgow, Scotland: William Collins, 1997.

Price, Huw. *Time's Arrow and Archimedes' Point: New Directions for the Physics of Time*. New York: Oxford University Press, 1997.

Randers, Jørgen. *2052: A Global Forecast for the Next Forty Years: A Report to the Club of Rome Commemorating the 40th Anniversary of The Limits to Growth*. White River Junction, VT: Chelsea Green Publishing, 2012.

Raphals, Lisa. *Divination and Prediction in Early China and Ancient Greece*. Cambridge: Cambridge University Press, 2013.

Raskin, Paul. *Journey to Earthland: The Great Transition to Planetary Civilization*. Boston: Tellus Institute, 2016.

Raworth, Kate. *Doughnut Economics: Seven Ways to Think Like a 21st-Century Economist*. London: Penguin Random House, 2017.

Rawson, Elizabeth. *Cicero: A Portrait*. Bristol Classical Paperbacks. 1975. Bristol: Bristol Classical Press, 2001, based on the 1985 edition.

Redmond, Geoffrey. *The I Ching (Book of Changes): A Critical Translation of the Ancient Text*. London: Bloomsbury, 2017.

Rees, Martin. *Just Six Numbers: The Deep Forces that Shape the Universe.* New York: Basic Books, 2000.

————. *On the Future: Prospects for Humanity.* Princeton, NJ: Princeton University Press, 2018.

Rescher, Nicholas. *Predicting the Future: An Introduction to the Theory of Forecasting.* Albany: State University of New York Press, 1998.

————. "Predicting and Knowability: The Problem of Future Knowledge." In *The Limits of Science,* vol. 109, Poznan Studies in the Philosophy of Humanities and the Sciences, edited by W. J. Gonzalez, 115–33. Leiden, Netherlands: Brill, 2016.

Riahi, Keywan, Detlef P. van Vuuren, Elmar Kriegler, Jae Edmonds, Brian C. O'Neill, Shinichiro Fujimori, Nico Bauer, et al., eds. "The Shared Socioeconomic Pathways and Their Energy, Land Use, and Greenhouse Gas Emissions Implications: An Overview." *Global Environmental Change* 42 (2017): 153–68.

Richerson, Peter J. "An Integrated Bayesian Theory of Phenotypic Flexibility." *Behavioral Processes* 161 (2019): 54–64.

Richerson, Peter J., Robert Boyd, and Robert L. Bettinger. "Was Agriculture Impossible during the Pleistocene but Mandatory during the Holocene? A Climate Change Hypothesis." *American Antiquity* 66, no. 3 (2001): 387–411.

Riggs, Peter. "Contemporary Concepts of Time in Western Science and Philosophy." In McGrath and Jebb, *Long History, Deep Time,* 47–66.

Robinson, Kim Stanley. "The Realism of Our Times: Kim Stanley Robinson on How Science Fiction Works." Interview with John Plotz, *Public Books,* September 23, 2020, https://www.publicbooks.org/the-realism-of-our-times-kim-stanley-robinson-on-how-science-fiction-works/.

————. *Red Mars, Blue Mars, Green Mars.* New York: Bantam Spectra, 1992–96.

Rockström, Johan, and Mattias Klum. *Big World: Small Planet.* Stockholm: Max Ström Publishing, 2015.

Rorvig, Mordechari. "How to Spot an Alien Megastructure." *New Scientist,* January 30, 2021, 45–47.

Rose, Deborah Bird. *Dingo Makes Us Human: Life and Land in an Australian Aboriginal Culture.* Cambridge: Cambridge University Press, 2000.

Rosenbaum, S. "100 Years of Heights and Weights." *Journal of the Royal Statistical Society. Series A* (Statistics in Society) 151, no. 2 (1988): 276–309.

Rosling, Hans, and Ola Rosling. *Factfulness: Ten Reasons We're Wrong about the World — and Why Things Are Better Than You Think.* London: Sceptre, 2018.

Roth, Gerhard. *The Long Evolution of Brains and Minds.* New York: Springer, 2013.

Russell, Bertrand. *History of Western Philosophy.* 2nd ed. London: Unwin Paperbacks, 1979.

————. "Psychological and Physical Causal Laws." In *Basic Writings,* 288 (from *The Analysis of Mind,* London: Allen & Unwin; New York: Macmillan, 1921).

Russell, Stuart. *Human Compatible: AI and the Problem of Control.* New York: Penguin, 2019.

Ryan, W. F. *The Bathhouse at Midnight: An Historical Survey of Magic and Divination in Russia.* University Park: Pennsylvania State University Press, 1999.

Rynasiewicz, Robert. "Newton's Views on Space, Time, and Motion." *The Stanford Encyclopedia of Philosophy.* http://plato.stanford.edu/archives/fall2008/entries/newtonstm/.

Sabrin, Kaeser M., et al. "The Hourglass Organization of the C. elegans Connectome." *BioRxiv: The Preprint service for Biology,* April 5, 2019, https://www.biorxiv.org/content/10.1101/600999v2.

Safina, Carl. *Becoming Wild: How Animal Cultures Raise Families, Create Beauty, and Achieve Peace.* New York: Henry Holt, 2020. My thanks to Rida Vaquas for this reference.

Sagan, Carl. *The Cosmic Connection: An Extraterrestrial Perspective.* Cambridge: Cambridge University Press, 2000.

Sahlins, Marshal. "The Original Affluent Society." In *Stone Age Economics,* 1–39. London: Tavistock, 1974.

Sardar, Ziauddin. *Future: All That Matters.* London: John Murray, 2013.

Sargent, Lyman Tower. *Utopianism: A Very Short Introduction.* Oxford: Oxford University Press, 2010.

Scheidel, Walter. *The Great Leveler: Violence and the Global History of Inequality from the Stone Age to the Present.* Princeton, NJ: Princeton University Press, 2018.

Schrödinger, Erwin. *What Is Life?* 1944. Cambridge: Cambridge University Press, 2000.

Schroeter, John, ed. *After Shock: The World's Foremost Futurists Reflect on 50 Years of Future Shock — and Look Ahead to the Next 50.* Bainbridge Island, WA: Abundant World Institute, 2020.

Schwab, Klaus, with Peter Vanham. *Stakeholder Capitalism: A Global Economy That Works for Progress, People and Planet.* Hoboken, NJ: Wiley, 2021.

Schwartz, Peter. *The Art of the Long View: Planning for the Future.* Sydney: Currency Paperback, 1996.

Seth, Anil. *Being You: A New Science of Consciousness.* London: Faber & Faber, 2021.

Shah, Karina. "Complex Life's Days Are Numbered." *New Scientist,* March 6, 2021, 12.

Shapin, Steven. *The Scientific Revolution.* Chicago: University of Chicago Press, 1996.

Sheldrake, Merlin. *Entangled Life: How Fungi Make Our Worlds, Change Our Minds, and Shape Our Futures.* New York: Random House, 2020.

Shostak, Seth. "The Value of 'L,'" in Dick and Lupisella, *Cosmos & Culture: Cultural Evolution in a Cosmic Context,* 399–414.

Silver, Nate. *The Signal and the Noise: The Art and Science of Prediction.* London: Penguin, 2012.

Simard, Suzanne. *Finding the Mother Tree: Uncovering the Wisdom and Intelligence of the Forest.* New York: Penguin, 2021.

Bibliography of Cited Sources

Sinclair, David A., and Matthew D. LaPlante. *Lifespan: Why We Age — and Why We Don't Have To.* New York: Atria, 2019.

Slaughter, Richard A., ed. *Knowledge Base of Futures Studies [KBFS].* Hawthorn, Australia: DDM Media Group, 1996. CD-ROM Professional ed., 2005.

Smil, Vaclav. *Harvesting the Biosphere: What We Have Taken from Nature.* Cambridge, MA: MIT Press, 2013.

————. *Numbers Don't Lie: 71 Things You Need to Know about the World.* New York: Penguin, 2020.

Smolin, Lee. *The Life of the Cosmos.* London: Phoenix, 1998.

Sornette, Didier. "Dragon-kings, Black Swans, and the Prediction of Crises." *International Journal of Terraspace Science and Engineering* 2, no. 1 (2009): 1–18.

Srubar, Will. "Buildings Grown by Bacteria — New Research Is Finding Ways to Turn Cells into Mini-Factories for Materials." The Conversation, March 23, 2020. https://theconversation.com/buildings-grown-by-bacteria-new -research-is-finding-ways-to-turn-cells-into-mini-factories-for-materials -131279.

Stableford, Brian, and David Langford. *The Third Millennium: A History of the World, AD 2000–3000.* London: Sidgwick and Jackson, 1985.

Stapledon, Olaf. *Star Maker.* London: Methuen, 1937.

Steffen, Will, Wendy Broadgate, Lisa Deutsch, Owen Gaffney, and Cornelia Ludwig. "The Trajectory of the Anthropocene: The Great Acceleration." *Anthropocene Review* 2, no. 1 (2015): 81–98.

Stewart, Ian. *Do Dice Play God? The Mathematics of Uncertainty.* London: Profile, 2019.

Stewart, Randall. "The Sortes Barberinianae within the Tradition of Oracular Texts." Chap. 8 in Luijendijk and Klingshirn, *My Lots Are in Thy Hands.*

Strathern, Oona. *A Brief History of the Future.* London: Constable and Robinson, 2007.

Swain, Tony. *A Place for Strangers: Toward a History of Australian Aboriginal Being.* Melbourne: Cambridge University Press, 1993.

Swirski, Peter, ed. *A Stanislaw Lem Reader.* Evanston, IL: Northwestern University Press, 1997.

Szostak, Rick. *Making Sense of the Future.* New York: Routledge, 2022.

Taleb, Nassim Nicholas. *The Black Swan: The Impact of the Highly Improbable.* New York: Random House, 2007.

Tedlock, Barbara. "Toward a Theory of Divinatory Practice." *Anthropology of Consciousness* 17, no. 2 (2008): 62–77.

"Temporalities." Forum in *Past and Present,* no. 243 (2019).

Thomas, N., and C. Humphrey, eds. *Shamanism: History and the State.* Ann Arbor: University of Michigan Press, 1994.

Tomasello, Michael. *Why We Cooperate.* Cambridge, MA: MIT Press, 2009.

Toner, J. *Popular Culture in Ancient Rome.* Cambridge: Polity, 2009.

Toulmin, Stephen, and June Goodfield. *The Discovery of Time.* Chicago: University of Chicago Press, 1965.

Turner, G. M. "A Comparison of The Limits to Growth with 30 Years of Reality." *Global Environmental Change* 18, no. 3 (2008): 397–411. https://doi .org/10.1016/j.gloenvcha.2008.05.001.

———. "Is Global Collapse Imminent? An Updated Comparison of *The Limits to Growth* with Historical Data." MSSI Research Paper No. 4, Melbourne Sustainable Society Institute, University of Melbourne, 2014. https://sustainable.unimelb.edu.au/publications/research-papers/is -global-collapse-imminent.

United Nations Environment Programme. *Making Peace with Nature*. 2021. https://www.unep.org/events/unep-event/launch-unep-making-peace -nature-report.

Urry, John. *What Is the Future?* London: Polity, 2016.

Vitebsky, Piers. *The Shaman*. Basingstoke: Macmillan, 1995.

Vollset, Stein Emil, Emily Goren, Chun-Wei Yuan, Jackie Cao, Amanda E. Smith, Thomas Hsiao, Catherine Bisignano, et al. "Fertility, Mortality, Migration, and Population Scenarios for 195 Countries and Territories from 2017 to 2100: A Forecasting Analysis for the Global Burden of Disease Study." *Lancet* 396, no. 10258 (2020): 1285–1306. https://doi.org /10.1016/S0140-6736(20)30677-2.

Voros, Joseph. "Big Futures: Macrohistorical Perspectives on the Future of Humankind." In *The Ways That Big History Works: Cosmos, Life, Society and Our Future*. Vol. 3 of *From Big Bang to Galactic Civilizations: A Big History Anthology*, edited by Barry Rodrigue, Leonid Grinin, and Andrey Korotayev, 403–36. Delhi: Primus Books, 2017.

———. "Big History and Anticipation: Using Big History as a Framework for Global Foresight." In *Handbook of Anticipation: Theoretical and Applied Aspects of the Use of Future in Decision Making*, edited by R. Poli. Cham, Switzerland: Springer International, 2017. https://doi.org/10.1007/978-3-319-31737 -3_95-1.

———. "On the Philosophical Foundations of Futures Research." In *Knowing Tomorrow? How Science Deals with the Future*, edited by P. van der Duin, chap. 5, 69–90. Delft, the Netherlands: Eburon Academic Publishers, 2007.

Wagar, W. Warren. "H.G. Wells and the Genesis of Future Studies." In Slaughter, *Knowledge Base of Futures Studies*, vol. 1, pt. 1.

———. *A Short History of the Future*. 3rd ed. Chicago: University of Chicago Press, 1999.

Waldrop, M. Mitchell. *Complexity: The Emerging Science at the Edge of Order and Chaos*. 1992. New York: Open Road Media, 2019.

Walls, Jerry, ed. *The Oxford Handbook to Eschatology*. Oxford: Oxford University Press, 2010.

Watts, Duncan J. "The 'New' Science of Networks." *Annual Review of Sociology* 30 (2004): 243–70.

Weart, Spencer. "The Development of the Concept of Dangerous Anthropogenic Climate Change." In *The Oxford Handbook of Climate Change and*

Society, edited by John Dryzek, Richard B. Norgaard, and David Schlosberg, 67–81. Oxford: Oxford University Press, 2011.

Weaver, Warren. *Lady Luck.* Dover: Penguin, 1963.

Wells, H. G. "The Discovery of the Future." *Nature,* February 6, 1902, 326–31.

———. *The Outline of History.* New York: Macmillan, 1920.

———. *The Rights of Man; or, What Are We Fighting For?* 1940. London: Penguin, 2015, with an introduction by Ali Smith.

———. *The Time Machine.* 1895.

Westfall, Richard S. *The Life of Isaac Newton.* Cambridge: Cambridge University Press, 1993.

Whitehead, A. N. *Adventures of Ideas.* New York: Free Press, 1933.

Wilczek, Frank. *Fundamentals: Ten Keys to Reality.* New York: Penguin, 2021.

Wohlleben, Peter. *The Hidden Life of Trees: What They Feel, How They Communicate.* Greystone Books, 2016.

Wolchover, Natalie. "Does Time Really Flow? New Clues Come from a Century-Old Approach to Math." *Quanta Magazine,* April 7, 2020.

Wolpert, Lewis. *Developmental Biology: A Very Short Introduction.* Oxford: Oxford University Press, 2011.

Wood, Barry. "Big History and the Study of Time: The Underlying Temporalities of Big History." In Benjamin, Quaedackers, and Baker, *The Routledge Companion to Big History,* 37–56.

Woodburn, James. "Egalitarian Societies." *Man, the Journal of the Royal Anthropological Institute* 17, no. 3 (1982): 432–51.

Wooton, David. *The Invention of Science: A New History of the Scientific Revolution.* New York: Penguin, 2015.

Wordsworth, Jonathan, and Jessica Wordsworth, eds. *The Penguin Book of Romantic Poetry.* London: Penguin, 2003.

Zimmer, Carl. *Microcosm: E. Coli and the New Science of Life.* New York: Vintage, 2009.

Zinkina, Julia, Leonid Grinin, Ilya Ilyin, Alexey Andreev, Ivan Aleshkovskii, and Andrey Korotayev. *Big History of Globalization: From the Big Bang to Modernity.* Cham, Switzerland: Springer, 2018.

Index

Tables and figures are indicated by an italic *t* or *f* following the page number.

Index

Paul (Saint), 3
Pavlov, Ivan, 110
Pavlovian (classical) conditioning, 110
Pearl, Judea, 36, 191
Peccei, Aurelio, 213
Penzias, Arno, 295
Perlmutter, Saul, 297
permanence, in ideas about time, 128, 134, 136–139
Persian Achaemenid Empire, 148
perspectival nature of experience of time, 41–46
Petty, William, 201–202
pheromones, 94–95
"A Philosophical Essay on Probabilities" (Laplace), 30–31
philosophy of time
    approaches to time, 17–19
    arrow of time, 36–38
    A-series time, 19–23, 22f, 23f
    B-series time, 24–28, 24f
    causation, 29, 34–38
    determinism, 29–34
    overview, 15–17
photosynthesis, 91–92
Piketty, Thomas, 245
Pinker, Steven, 123
planetary futures, 286–293
planetary limits to growth, 226–231, 234, 237–246
planetary migrations, 273–277
planet management, 262–264.
    See also conscious planet
plants, future management by, 91–100
plate tectonics, 287
"plausible" domain, 63–64, 63f, 66
Plutarch, 111
poison oracles, 168–169
Polak, Fred, 213
politics
    in middle futures, 264
    in near futures, 219–220, 259
    and transition to conscious planet, 249

trends in, 234, 246–247
turbulence in imagined futures, 250–251
Popper, Karl, 125
popular future thinking in agrarian era
    conflicts between elite future thinking and, 149–152
    glimpses of, 166–174
    overview, 146–149
population
    in future scenarios, 257, 258f
    growth trends, 237–239
Porter, Roy, 189
"possible" domain, 63, 63f, 64, 66
The Power of Movement in Plants (Darwin), 99
practical future thinking. See relationship, time as
predetermination, 29–34
predictability, domains of, 62–65, 62t, 63f
prediction. See also trend hunting
    abandoning hopes of perfect, 32–33
    in modern era, 188, 206–214
    by oracles in agrarian era, 153–155
    retrospective, 111–112
    trend hunting and, 53–60
preference, domains of, 61f, 62
"preposterous" domain, 62–63, 63f
present
    in A-series time, 21, 22, 23f
    in B-series time, 24, 24f, 25–27
    ideas about in foundational era, 137–138
    trend hunting and, 59–60
Price, Huw, 25
Principia Mathematica (Newton), 19
prior, in Bayesian analysis, 72
probability. See also statistics
    Bayesian analysis, 72
    in modern era future thinking, 188, 191–201, 199f
"probable" domain, 63f, 64

About the Author

David Christian is a Distinguished Professor of History at Macquarie University and director of Macquarie University's Big History Institute. He cofounded the Big History Project with Bill Gates, his Coursera MOOCs are popular around the world, and he is cocreator of the Macquarie University Big History School. He has delivered keynotes at conferences around the world, including the Davos World Economic Forum, and his TED Talk has been viewed more than twelve million times. He is the author of numerous books and articles, as well as the *New York Times* bestseller *Origin Story*.